Bugs in the System

Bugs in the System

Redesigning the Pesticide Industry for Sustainable Agriculture

Edited by

William Vorley and Dennis Keeney

First published by Earthscan in the UK and USA in 1998

This edition published 2013 by Earthscan

For a full list of publications please contact:

Earthscan
2 Park Square, Milton Park, Abingdon, Oxon OX14 4RN
Simultaneously published in the USA and Canada by Earthscan
711 Third Avenue, New York, NY 10017, USA

First issued in hardback 2017

Earthscan is an imprint of the Taylor & Francis Group, an informa business

A catalogue record for this book is available from the British Library

ISBN 13: 978-1-8538-3429-5 (pbk)
ISBN 13: 978-1-1384-7160-3 (hbk)

Typesetting, page design and figures by PCS Mapping & DTP, Newcastle upon Tyne
Cover design by Dan Mercer

Contents

List of Figures, Tables and Boxes

Figures

Tables

BOXES

Acronyms and Abbreviations

ABL Adapting By Learning (Canada)
AI active ingredient
AIM Accuracy in Media
ARA Agricultural Retailers Association (US)
BCSD Business Council for Sustainable Development
BIOS Biologically Integrated Orchard Systems (US)
BSA British Standards Authority
BSE Bovine Spongiform Encephalopathy
Bt *Bacillus thuringiensis*
CAP Common Agricultual Policy
CAST Council for Agricultural Science and Technology (US)
CBI Confederation of British Industries
CDOFA California Department of Food and Agriculture
CEO chief executive officer
CERES Coalition for Environmentally Responsible Economics
CFC chlorofluorocarbon
CISA Community Involved in Sustainable Agriculture (US)
CJD Creutzfeld-Jakob Disease
CLM Centre for Agriculture and Environment (The Netherlands)
CPI Consumer Price Index
CSARE Consortium on Sustainable Agriculture Research and Extension
CSERGE Centre for Social and Economic Research on the Global Environment
DDT dichlorodiphenyltrichloroethane
DoE Department of the Environment (UK)
EFTEC Economics for the Environment Consultancy Ltd
EIL economic injury level
EIQ environmental impact quotient
EPA US Environmental Protection Agency
EU European Union
FAO United Nations Food and Agriculture Organization
GATT General Agreement on Tariffs and Trade
GNP gross national product
HFCS high-fructose corn syrup
HMSO Her Majesty's Stationery Office
HRM holistic resource management
IATP Institute for Agriculture and Trade Policy
ICC International Chamber of Commerce
IFPRI International Food Policy Research Institute
IIED International Institute for Environment and Development
INCOSE International Council On Systems Engineering
IPM integrated pest management
ISO International Standards Organization
MAFF Ministry of Agriculture, Fisheries and Food (UK)
MB marginal benefit
MB methyl bromide
MC marginal cost

MEC	marginal external cost
MFT	marriage and family therapy
MRL	maximum residue limit
MTC	maximum tolerable concentration
NAS	National Academy of Science
NCOSE	National Council on Systems Engineering
NGO	non-governmental organization
NRA	National Rivers Authority
NRDC	Natural Resources Defense Council (US)
OECD	Organization for Economic Cooperation and Development
OTA	Office of Technology Assessment (US)
PAC	political action committee
PCB	polychlorinated biphenyl
PIC	prior informed consent
PR	public relations
R&D	research and development
UN	United Nations
UNCED	United Nations Conference on Environment and Development
UNCTAD	United Nations Conference on Trade and Development
UNEP	United Nations Environment Programme
UNDP	United Nations Development Programme
USDA	US Department of Agriculture
VOSL	value of a statistical life
WCED	World Commission on Environment and Development
WHO	World Health Organization
WRI	World Resources Institute
WTA	willingness to accept
WTO	World Trade Organization
WTP	willingness to pay
WWF	World Wide Fund For Nature

CONTRIBUTORS

Sam Alessi holds a doctorate in soil physics and has extensive knowledge of the aerospace-derived discipline of systems engineering and systems architecture. He is a coprinciple of A&M Consulting. He is also the current president of the North Star Chapter of the International Council On Systems Engineering (INCOSE) and is cochair of the INCOSE systems engineering principles working group. Sam is working with a team as part of the Consortium on Sustainable Agriculture Research and Extension (CSARE) to enhance technology assessment of future social impact. He is the primary architect and programme manager of the farm-based management information system entitled FARMWIN and has worked to introduce customer-focused engineering approaches to agricultural research and industry.

Vivien Foster is a Senior Economic Consultant at Oxford Economic Research Associates Ltd, and is also affiliated with the Centre for Social and Economic Research on the Global Environment at University College London.

Beth Franklin is a part-time faculty member in the Faculty of Environmental Studies at York University, Ontario, and a principal of the ABL Group. She has a doctorate in human and organization systems. Professional and research interests include planning and management in multi-organizational domains, educational design, international development, and social change. She has a particular interest in using the 'future search' methodology in ways that reflect action-learning principles.

Dennis Keeney has been Director of the Leopold Center for Sustainable Agriculture in Iowa (USA) since September 1988. He was formerly chair of the Land Resources Graduate programme in the Institute for Environmental Studies, and earlier chair of the Soil Science Department at the University of Wisconsin–Madison. Among Dennis's awards is the American Society of Agronomy's Environmental Quality Research Award.

Dominik Koechlin has a directorate in Law from the University of Bern where his thesis was on the precautionary principle in Swiss environmental law, and an MBA from INSEAD, Fontainebleu. Together with Kaspar Mueller, he founded Ellipson – a Swiss-based consultancy specializing in environmentally conscious management and strategy – in 1990, and went on to coedit the book *Green Business Opportunities: the Profit Potential* (1992). Dominik has been a frequent speaker at environmental management workshops. In 1996 he moved from Ellipson to new position as director of corporate development at Swiss Telecom.

Mick Mayhew holds a doctorate in family systems therapy and has practised as a mental health clinician for ten years. He is coprinciple of A&M Consulting. Mick has designed, developed and overseen many successful programmes within mental health and research communities. He is also a member of INCOSE. He has delivered workshops and presentations to aerospace engineers, agricultural engineers, scientists and practitioners, mental health providers, civic organizations and religious communities.

David Morley is Professor in Environmental Studies at York University, Ontario, and a founder member of York's ABL Group. He has been involved in many projects that focus on the application of action learning to domain-based organizational change. He has coedited (with Susan Wright) *Learning Works: The Search for Organizational Futures* (1989).

Susana Mourato is Research Fellow at the Centre for Social and Economic Research on the Global Environment, London, UK, and Lecturer in Economics at Universidade Católica Portuguesa, Lisbon, Portugal.

Ece Özdemiroglu is director of Economics For The Environment Consultancy Ltd (EFTEC). She is a specialist in project appraisal and monetary valuation related to environmental issues. Ece has performed consulting work for the World Bank, various UN organizations, the European Commission, UNCTAD, the Asian Development Bank, the UK Department of the Environment, Transport and Regions and the UK National Rivers Authority (NRA).

David Pearce is Professor of Environmental Economics at University College London and Associate Director of the Centre for Social and Economic Research on the Global Environment (CSERGE; University College London and University of East Anglia, UK). He has worked for the OECD, UNDP, UNEP, the World Bank, and is author and editor of 50 books including the *Blueprint for a Green Economy* series (1989, 1991, 1993, 1995, 1996) and *World Without End: Environment, Economics and Sustainable Development* (1993). He has been Special Advisor to the UK Secretary of State for the Environment and the recipient of the United Nations Global 500 award for services to the world environment. He is presently chairman of the UN Economic Commission for Europe's Economics Group on Acid Rain.

Jules Pretty is Director of the Centre for Environment and Society at the University of Essex, UK. He is also Research Fellow at the International Institute for Environment and Development (IIED), where he was Director of the Sustainable Agriculture Programme from 1989 to 1997. He has published widely, his books including *The Living Land* (in press), *Regenerating Agriculture: Policies and Practice for Sustainability and Self-Reliance* (1995) and *Unwelcome Harvest: Agriculture and Pollution* (1991, coauthored). He is a founding member of the Agricultural Reform Group, and a trustee and advisor to a range of private and public organizations.

Robert Tinch is an environmental economist currently reading for a PhD in the Environment Department at the University of York, UK, researching the links between the complexity and the stability of ecological–economic models. He is a director of EFTEC-York.

William Vorley is coordinator of the project which has culminated in this book. He has a doctorate in applied insect ecology and field research experience in England, Japan, Malaysia and Indonesia. After nine years working with Ciba-Geigy in Asia and at their Swiss headquarters, he resigned to take up the position of Visiting Scientist at the Leopold Center for Sustainable Agriculture. His commercial experience covered both new pesticide development and extension. His successful 'Farmer Support' campaign to look for business opportunities from building small farmers' skills in safety and integrated pest management put Ciba-Geigy at the forefront of the service-oriented agrochemical business in developing countries. William is currently Director of the Environment and Agriculture Programme at the Institute for Agriculture and Trade Policy (IATP) in Minneapolis, US.

Anja Wittke is Senior Consultant at Ellipson AG in Basle, Switzerland, with a special focus on sustainable controlling and accounting concepts. She has a Masters degree in business administration from the University of Muenster, Germany. She has taken part in various environmental management projects at the Institute for Environmental Research in Berlin, and was responsible for the development, organization and implementation of a one year strategic management, controlling and marketing training and education programme for middle management in the new German provinces.

Preface

The images around the pesticide issue are startlingly vivid. A spectre of a silent spring without birdsong; slum dwellers in Bhopal gasping for air; airplanes flying low over fields, trailing swaths of spray; graphs of a Malthusian human population boom; peasants helplessly watching pests steal their crop and their livelihood; a proud American farmer and family standing against the sunset in front of silos full of grain. Small wonder that the debate about the role of pesticides and pesticide manufacturers in a more sustainable agriculture gets hopelessly lost in a quagmire of apparently irreconcilable certainties and stereotypes.

An investigation of potential routes out of this quagmire may seem quixotic. Certainly, it is impossible to take a balanced view of the issues; one person's balance is another person's bias. But the industry's own language on the greening of business and sustainable development seems a good place to start.

The chemical industry's language about corporate environmental responsibility and sustainable development evolved into something anticipatory and relatively radical in the late 1980s and early 1990s, partly as a build up to the United Nations Conference on Environment and Development (UNCED or 'Earth Summit') in 1992, and partly as a result of the disasters at Bhopal, Basle and Prince William Sound. This new language of 'equal ranking economic, environmental and social responsibilities', of 'internalizing external costs', of 'new business in sustainable development' and of 'new collaborations with all company stakeholders' has a risk, of course, of turning into greenwash if society sees only token gestures which tweak the margins of the status quo.

Such a risk became apparent to one of the authors, whose real-life insider's experience at the headquarters of the world's largest pesticide company was the inspiration for this book. That company's vision for sustainable development created a culture in which tremendous improvements in product stewardship could be achieved. The industry's most ambitious skills-building programme ever attempted in developing countries was approved, funded and staffed, working on the premise that pesticide overuse and misuse is bad news in the long term not just for farmers, but also for pesticide manufacturers. However, as this exercise in implementing a company's vision progressed, the strains and contradictions inherent in that vision became clear. There was much talk about sustainable agriculture, but commitments to economic, environmental and social responsibilities appeared to many employees to become only of equal ranking in the Orwellian sense; some were more equal than others.

Clearly, if companies were going to be serious about sustainable agriculture, there was a need for a comprehensive and detached analysis of the industry, looking behind the claims and counter-claims, and viewing the issues and opportunities from many angles. Hence this book, which is a deliberate multidisciplinary approach to lay the groundwork of constructive options for this most controversial of industries, grounded in industry's own language and own commitments.

The project was coordinated from the Leopold Center for Sustainable Agriculture in the US state of Iowa. Iowa is the agricultural heartland of North America. The rolling landscape is a testament to agricultural efficiency – maize and soybeans dominate the landscape and, thanks to the natural fertility of the deep prairie soil, an ideal weather pattern and modern plant varieties, fertilizers and pesticides can be used to their utmost effectiveness to achieve the world's highest sustained yields of these crops. But this abundance comes at a cost, and water quality, soil health, wildlife abundance and diversity, farm labour, and the rural economy have all witnessed dramatic declines in parallel with the intensification and specialization of farming. The presence of herbicides and nitrate in ground water and drinking water rang alarm bells, to the extent that – through a piece of landmark legislation – Iowa lawmakers established the Leopold Center to drive research and extension programmes on alternative, more regenerative farming systems.

Early in 1994, we contracted with research groups in London and Switzerland who could take the greening of business language and turn it into a rigorous analysis with minimum bias. These groups were charged with asking some key questions. What are the drivers of pesticide use, and what is the connection between pesticide use, feeding the world and preserving agriculture's natural resource base? What are the 'real costs' of pesticide use, and how would that affect the relative sustainability of pesticide-based farming systems relative to other methods? What new business opportunities exist for the pesticide industry in a more regenerative agriculture, and how prepared is the industry to exploit those potential new areas? As the project progressed, we brought in two more groups, from the US and Canada, to explore the implications of industry's commitments to broader stakeholder involvement in strategy and technology choice.

While the project was unfolding and as the book was being written, parts of the pesticide industry were embarked on a rapid, unprecedented change, moving from offshoots of chemical giants to focused 'bioscience' companies, with agricultural divisions investing heavily in biotechnology research and access to seed markets. The substitution of engineered seeds for chemical pesticides is itself being hailed as a strategy for sustainability, and we discuss the merits of this claim with the benefit of hindsight over the chemical pesticide era.

William Vorley and Dennis Keeney
Minneapolis, Minnesota, and Ame, Iowa
October 1997

ACKNOWLEDGEMENTS

We greatly appreciate the contributions from the project partners between 1994 and 1997, and for their chapters presented in this book. Much of this work extended well beyond the call of duty.

The work was funded in part by the Iowa Department of Agriculture and Land Stewardship (IDALS) through a generous grant from the US Environmental Protection Agency's (EPA) Environmental Technology Initiative. The work presented here does not, of course, necessarily represent the views of these sponsors. We acknowledge the invaluable assistance of EPA Region 7 (Joe Cothern) and IDALS (Jim Ellerhoff) as well as Iowa State University.

We are also deeply grateful to the staff of the Leopold Center – especially April Franksain and Ken Anderson – and the Center's Advisory Board, for their outstanding work and facilitation.

Numerous farmer, scientific and NGO contacts helped to develop the ideas laid out in the chapters of the book. Groups of volunteers from all sectors of society comprised the focus groups at the core of Chapter 6 and the Future Search conference described in Chapter 7; the authors and editors acknowledge their tremendous contributions with thanks. Also, the Future Search depended on the small group facilitation of Clyde Boysen, Mick Mayhew and Laurie Schmidt, and the search was organized by the Institute for Agriculture and Trade Policy in Minneapolis, with the invaluable assistance of Mark Ritchie, Jim Kleinschmit and Cali Brooks.

We would also like to express our appreciation to reviewers, including Dan Zinkand, Diane Mayerfeld and Mary Adams, as well as the staff of the World Resources Institute (WRI), the World Wide Fund For Nature (WWF) and the Henry A Wallace Institute for Alternative Agriculture. From the initial contact with Jo O'Driscoll through to publication, Earthscan Publications Ltd have been patient and enthusiastic. Thank you.

This book is dedicated to our families, with love.

1

The Greening of Industry versus Greenwash: Introducing a Case Study

William Vorley and Dennis Keeney

Progress towards sustainable development makes good business sense because it can create competitive advantages and new opportunities.

Declaration of the Business Council for Sustainable Development (Schmidheiny & BCSD, 1992, 9 pxii)

What looks like progress is very often simply the immune system of the status quo defending itself.

David Edwards, 1995, p188

Bugs in the System: Reason for the Study

How does a company objectively review its role in contributing to, or detracting from, a sustainable future? How does it recognize opportunities in sustainable development beyond public relations 'greenwash'? For established industries, this subject area is something of a conceptual vacuum. There is a shortage of case studies which look behind the claims and counter-claims to view the issues and opportunities from many angles.

The pesticide industry is a fascinating model for a case study. The industry has some leading CEOs whose endorsements of sustainable development verge on the evangelical. The industry has, through baptism by fire, developed some rather radical initiatives for environmental protection and public accountability, such as the Responsible Care programme. The leaders of three of the largest pesticide producers – Ciba-Geigy (now Novartis), DuPont and Dow – sat on the Business Council for Sustainable Development whose 1992 declaration is seen as one of the most radical statements by business leaders on the need for fundamental change. Its products are designed to save the staff of life – food – from destruction by the ravages of insects, weeds and diseases. But those very products are one of the most controversial components of agriculture, synonymous in the eyes of many with the fundamental *un*sustainability of modern farming.

The challenges and contradictions do not end there. When 'sustainable' industries turn to cradle-to-grave product stewardship to avoid the release of harmful chemicals into the environment, pesticide manufacturers are faced with the fact that their products are deliberately released into the environment. From the part of the product's life-cycle, which lies outside the factory gates, only an empty container remains for recycling. And while 'sustainable' industries aim to preserve biodiversity, pesticides are intended to reduce biodiversity, in order to tilt the balance in favour of crops and humans rather than to encourage the diversity of pests which compete for those crops. Furthermore, the industry carries a heavy burden of history of producing chemicals which still today, long after their legislative demise in most countries, are moving around the world and accumulating in the bodies of humans and wildlife. Some of the great watersheds of concern over environmental health and corporate responsibility have been related to pesticides and still echo in the public consciousness – DDT and Rachel Carson's *Silent Spring*, Agent Orange, Bhopal and the Sandoz fire in Basle that polluted the Rhine River. The industry stands accused of defending products against a tide of evidence of harm to human and ecosystem health, of vehemently opposing regulatory reforms and moves to reduce reliance on pesticides, and of double standards between the developed and developing worlds.

There is also the challenge of timing. The late 1990s, when companies should be implementing their pledges of 'far-reaching shifts in corporate attitudes and new ways of doing business', have turned out to be a far leaner and meaner era than the more visionary late 1980s and early 1990s. The pesticide industry is experiencing a wave of consolidation, cost-cutting and layoffs in order to maintain profitability. Sustainable development seems to have taken a back seat for many companies, while survival in this maturing industry takes precedence. Such conditions are considered far from ideal for the greening of corporate strategy (Walley and Whitehead, 1994). However, perhaps the most outstanding feature of the pesticide industry which makes for such an interesting case study is the dangerous orthodoxy which characterises the debate. The pesticide issue has attracted the full force of both apocalyptic environmentalism and the 'wise use' revisionism of commentators such as Gregg Easterbrook (1995) and Dennis Avery (1995). Pesticides are presented as either a cornerstone of efforts to feed the world, or a threat to planetary health. To give an example of the depth of this feeling, one letter we received stated bluntly that:

> *It does not take a book to describe how to redesign the pesticide industry. If the chemical companies had to pay for the lives they have destroyed and the damage to the earth they have caused, they would already be out of business.*

Constructive discussion on the future for pesticide producers in a more sustainable world – and the innovations in products and services which would be required for a more sustainable food and fibre production – is drowned out by the din of present information crossfire between deeply entrenched camps loyal to apparently irreconcilable paradigms, armed and supplied by some powerful vested interests. The proxy war has spread to university campuses in the form of sponsored research, contaminating this remaining bastion of scientific objectivity.

Orthodoxy breeds defensive, linear thinking and intellectual inbreeding, which in turn causes actors – businesses, regulators, activists – to push harder and harder on familiar solutions. The public become sceptical about the safety of their food, so industry launches a new information campaign to nurse them out of their illogical misconceptions. Governments find new conflict-laden areas for imposing costly regulations to limit pollution. The debate centres on ridiculous choices and grand dichotomies. Bugs or people? Technology: friend or foe? High-input agriculture or starvation? High-input agriculture or wildlife?

Readers familiar with systems theory will recognize here the characteristics of dysfunctional approaches to a complex social system (Senge, 1990, pp 57–67). Pushing harder with familiar solutions that do not work, either-or dilemmas, blaming outside circumstances and dualistic us-versus-them views are classic symptoms. There is a fascinating parallel with the dysfunctional way in which agriculture has approached the control of pests with pesticides in complex ecological systems. Direct war on pesticides with chemicals has drawn farming into a treadmill of resistant pests, new secondary pests and simplified ecologies which demand an even harder push with more expensive and sophisticated weapons from the chemists and gene engineers. From its own single perspective, the industry sees no options other than to further crank up the PR machine and dig deeper into the technology tool box until the wars against the industry's detractors and the wars against pests are won. Such attempts to deal with a situation by searching for what company managers can understand and cope with in terms of its existing paradigm are a symptom of strategic drift. Johnson and Scholes (1993, p63) commonly observe this tendency of searching for the familiar in the face of ambiguity and uncertainty in organizations with a high degree of homogeneity in the beliefs and assumptions which comprise it. But perhaps with divergent, constructive perspectives, we have the chance to find innovative solutions more in keeping with industry's sustainability talk. In his classic book *The Fifth Discipline*, Peter Senge (1990, pp 64–65) states that 'small, well-focused actions can produce significant, enduring improvements, if they're in the right place'. 'The only problem', adds Senge, 'is that high-leverage changes are usually counter-intuitive and nonobvious to most participants in the system... until we understand the forces at play in those systems.'

The purpose of this book is to gain an understanding of the system – the triangle of business, agriculture, and sustainability – in order to find those potential leverage points for change. Our null hypothesis – which we will revisit in Chapter 8 – is taken from the Business Council for Sustainable Development's 1992 declaration (Schmidheiny and BCSD, 1992, pp xi–xiii). It states that: 'Progress towards sustainable development makes good business sense because it can create competitive advantages and new opportunities.' The book records the research of five teams in North America and Europe that were invited to explore different perspectives, building on industry's own language, to improve our understanding and construct a road map of directions towards these new opportunities.

Consistent with the BCSD declaration, and the title of this book, we are looking beyond efficiency towards redesign (see Table 1.1). If the controversy over pesticides in sustainable agriculture were ended by cleaner pesticide production or recyclable containers, or the controversy over the impact of the automobile were resolved by catalytic converters, air bags and recyclable components, we could rely on management approaches for industry to meet their obligations. But as Schmidheiny and the BCSD (1992, 7) concede, 'Sustainable development will obviously require more than pollution prevention and tinkering with environmental regulations'; instead, 'profound changes in the goals and assumptions that drive corporate attitudes' are needed.

Our approach can be summarized as follows:

Table 1.1 *The distinction between efficiency, substitution and redesign*

	Efficiency	*Substitution*	*Redesign*
Company focus	Production	Products/Services	Company mission
Objective	Efficiency: fewer 'bads'	Effectiveness: more 'goods'	
Responsible parties	Engineers	R&D	All employees
Scope of sustainability	Environmental/technical		Social, ecological, economic
Catchwords	Eco-efficiency Win–Win Shareholder value Life-cycle analysis Industrial ecology		Deep design The Natural Step
Key business documents	Responsible Care Keidanren Charter ICC Declaration ISO 14000	CERES Principles BCSD Declaration	

Source: based on the concepts of Hill and MacRae (1995) and MacRae et al (1993)

- explore the context and understand the forces at play;
- look for high leverage points;
- target redesign, not efficiency;
- build on industry commitments;
- take an interdisciplinary approach.

Understanding the Forces at Play

The vision thing: Business goes sustainable

The controversial natures of pesticide manufacturing and chemical production are closely interlinked. The industry which releases more toxic substances into the environment than any other sector, which brought us CFCs, PCBs, Love Canal and Seveso, is also the industry which supplies the world with pesticides. Unpopular with the public and spurned by 'socially responsible' investment funds, the chemical industry is seen as the antithesis of ecologically based commerce in the same way as pesticides are seen by some commentators as the antithesis of ecologically based sustainable agriculture. Paul Hawken, himself a successful businessman, does not pull his punches:

> *If DuPont, Monsanto and Dow believe they are in the synthetic chemical production business, and cannot change this belief, they and we are in trouble. If they believe they are in business to serve people, to help solve problems, to use and employ the ingenuity of their workers to improve the lives of people around them by learning from the nature that gives us life, we have a chance.*
>
> (Hawken, 1993, pp 54–55)

Throughout the 1980s, the chemical industry responded to this criticism with the rhetoric of denial or neutralization, implying that they were being unfairly blamed by an irrational and misguided public (Tombs, 1993).

But in the late 1980s and early 1990s, especially in the build-up to the Rio Earth Summit, many CEOs of large multinational corporations became much more proactive and outspoken visionaries. Alex Krauer and Heini Lippuner of Ciba-Geigy kicked off Vision 2000, which spoke of living up to three equal-ranking responsibilities – economic success, social responsibility and responsibility for the environment. These visions were a response to the rift which had opened up between industry, especially the chemical industry, and society, and the rift between the perceived goals of industry and employees' personal values. They were a response to deep global societal concerns that environmental problems were getting worse, and that business and industry cared more about growth than about protecting the environment (Bloom, 1995). They were a response to the disasters in Bhopal, in Prince William Sound and in Basle. In a new spirit of frankness and openness, some CEOs questioned the wisdom of the frenzy of unbridled growth and of producing agricultural surpluses with the aid of chemicals. They called for a social consensus on the application of new technology and for the application of the polluter pays principle. They called for environmental protection standards 'set so high that enlightened firms with a talent for innovation and a willingness to accept risks would be motivated to surpass them if only for the competitive advantages they stand to gain in the process' (Krauer, 1992). Fifty business leaders in the Business Council for Sustainable Development spoke of the need for 'far-reaching shifts in corporate attitudes and new ways of doing business'.

By the mid 1990s, many companies – spurred on by the publication of the quantity of waste production – had made good progress with technocratic ecoefficiency – clean production, waste reduction and environmental reporting (see Saunders and McGovern, 1993, pp 191–202). But corporate sustainable development had hit hard times. The white-knuckle days of the mid 1990s were about maintaining profitability and market share in a tough maturing and globalizing market. Talk of an end to the period of quantitative growth in the chemical industry started to look premature as mergers and acquisitions continued apace. The efforts at top-down visioning began to look like a thin gloss overlaying existing corporate structures and job descriptions. Not much rethinking of corporate goals and strategy was taking place beyond the uncontroversial win–wins of pollution prevention. Discussion of the role of the industry's products in a sustainable world hardly saw the light of day. Keeping a job and sustaining shareholder value were the preferred hard alternatives to woolly sustainability talk. Each business sector worked hard – each with their own interpretation of their vision – even though the actions of some sectors were distinctly contradictory in terms of the overall corporate vision. The researchers discovered. The salesmen sold. The 'communicators' told the industry's side of the story. The lobbyists lobbied. Life was going on as if corporate vision had been a brief *affaire* of the late 1980s and early 1990s, the stuff of management seminars and expensive consultants.

The 'vision thing' seems to be caught in a classic social trap – a situation in which decisions based on short-term goals and narrow interests lead to long-term outcomes desired by no one (Constanza, 1988), except perhaps the shareholders and the Cassandras who had scoffed at the 'vision thing' since its inception. To understand the forces at play here, we must understand the context of industry's perspectives and industry's predicaments. This is the objective of Dominik Koechlin and Anja Wittke in Chapter 5 of this book.

The controversial nature of sustainable agriculture

In the eyes of most advocates of a more sustainable agriculture, pesticides symbolize everything which is unsustainable about modern agriculture: aggressive intervention with biocidal chemicals that endanger human health, disrupt ecological balance and drain cash from the farm economy, all in the name of efficiency; cheap food at a high price. For many specialists, a reduction in pesticide inputs is by definition an axiom of increased agricultural sustainability (Hansen, 1996).

Stereotyping and portraying issues in terms of black or white, unsustainable or sustainable, is endemic to the debate over agriculture. Sustainable agriculture is typically defined by contrasting it to conventional agriculture – a construct of all that is perceived to be unsustainable in the modern agri-food system (reductionism, technological intensity, domination of nature, exploitation, dependence, competition, etc), though this conventional agriculture is probably not an accurate representation of most mainstream farmers (Beus and Dunlap, 1991). Sustainable agriculture has thus gained an unfortunate reputation of pointing an accusing finger at the faults of mainstream farming and its suppliers. It is small wonder that the pesticide industry has reacted so defensively to the ascendance of sustainable agriculture on the international stage. Small wonder also that the farmers of the world have generally aligned themselves with the pesticide industry – who are at least producing something of use to them – rather than the opponents of pesticides. Farmers, have, after all, received scant advice on how to reduce their dependence on pesticides, other than to turn to genetically engineered crop varieties, and few government incentives to help make the transition. And the market is only just starting, in a few European countries, to give strong signals that consumers want to put their money where their mouth is; the bulk of consumers seem ready to set their suspicions about pesticides aside once inside the supermarket doors.

Industry definitions of sustainable agriculture have become reactive endorsements of the status quo. Conventional agriculture is sustainable agriculture. Business managers are searching for what they can understand and their understanding is that agriculture must sustain growing populations with cheap food and fibre through continued improvements in productivity, and that 'without chemicals, you will have to feed pests and people' (Ciba-Geigy, 1993). So we are confronted with industry definitions grounded in productivism; DowElanco (1994) states that: 'Sustainable farmers optimize inputs (financial investments) to maximize profits' and BASF have defined sustainable agriculture as 'that level of productivity that allows the agricultural enterprise to be economically competitive'.

These definitions, while remarkable in the narrowness of their vision, only serve to highlight the degree to which the concept of sustainability is contested when applied to either development or agriculture. Definitions may interpret sustainability as an ideology, a set of strategies or a set of goals (Hansen, 1996). They may be grounded in fundamentally different moral theories, from biocentrism to anthropocentric productivism (see Turner et al, 1994, p31). Such a huge vista allows industry to get away with viewing sustainable development through the prism of its own world view (Buttel, 1992). Definitions are therefore particularly slippery areas of common ground, and a poor foundation to build a study of redesigning an industrial sector. If sustainable is continually seen as synonymous with efficient or productive, how do we catalyse the far-reaching shifts in corporate attitudes and new ways of doing business, which business leaders say are required to create competitive advantages and new opportunities? How do we get sustainability out of the hands of the issue managers, production engineers and public relations departments, and into the hands of the corporate strategists?

THE REFERENCE POINTS

The first step is to untangle the rhetoric and counter-claims, and put the products of the pesticide industry firmly into the context of sustainability. Is industry's demographic imperative correct, or are there other forces driving agriculture's dependence on pesticides? What is being sustained with pesticides, for whose benefit and for how long? The analysis of these questions is the task of Pretty et al in Chapter 2. We can also look deeper into industry's own language and international agreements on sustainable development for reference points.

Business declarations and international agreements

The international agreements relevant to the role of pesticide manufacturers in sustainable development mainly revolve around the 1992 UNCED Earth Summit agreements, though, in the event, the activities of business received rather scant attention. The binding though legally unenforceable Rio Declaration, intended as a charter of basic principles for sustainable development, has all the signs of UN-style lowest common denominator negotiation (Thompson, 1993); nevertheless it does contain the first international statements accepting the polluter pays principle. The action plan for the Rio Declaration was spelled out in an 800-page non-binding set of guidelines called Agenda 21.

Agenda 21 is first and foremost an action plan for governments, as the architects and builders of sustainable development. The section on agriculture and natural resources makes numerous demands on governments, while not one mention of private sector involvement is included. References to the role of business and industry in Chapter 8 of Agenda 21 focus more on technology transfer and stewardship – avoiding harm – rather than trail-blazing the road to sustainability.

Stewardship is also the focus of a number of business initiatives on the environment, such as the Chemical Manufacturers Association's famous Responsible Care programme or the Japanese Keidanren's 1991 Global Environment Charter (see Table 1.2). But other declarations presume a powerful and proactive role for business at the helm of sustainable development, confident that 'the world is moving toward deregulation, private initiatives, and global markets'. The most radical of these which has pesticide producers as signatories is the Business Council for Sustainable Development (BCSD) 1992 declaration, published in the run up to the Rio Earth Summit. The three-page BCSD declaration is printed at the beginning of the book *Changing Course*, and although the whole 374 pages are ostensibly coauthored by the Swiss industrialist Stephan Schmidheiny and the BCSD, we should look to the declaration for the voices of BCSD members – who included the CEOs of Dow Chemical, DuPont and Ciba-Geigy – rather than the largely ghost-written (but more radical) book contents. We should note that the BCSD has not impressed everyone with their declaration and book. Greenpeace list the group in their *Guide to Anti-Environmental Organizations* (Deal, 1993), accusing them of 'a key role in watering down [Rio] treaties on biodiversity and global warming' (see also Willers, 1994).

The BCSD declaration and Agenda 21 have very different outlooks on the relative importance of the private and public sectors in achieving sustainable development. The signatories of the BCSD declaration maintained that it was in business's own interest to assume the reins of responsibility in a less regulated global free market. But there are two issues on which there is remarkable unanimity between the approaches – the correction of prices to reflect the unpaid environmental costs of their production and

Table 1.2 *Some business charters for the environment and sustainable development*

Organization	*Statement*	*Focus*
US Chemical Manufacturers Association (Canada: 1985)	Responsible Care (1988)	Safe production and handling Public information
Coalition for Environmentally Responsible Economies (CERES, US)	CERES Principles (formerly Valdez Principles, 1989)	Protection/restoration of biosphere Waste and risk reduction/ elimination Safe products and services Public disclosure Continuous improvement
Keidanren (Japan Federation of Economic Organizations)	Global Environment Charter (1991)	Innovative technologies Public information and community relations
International Chamber of Commerce (ICC)	ICC Business Charter for Sustainable Development (1991)	Employee and customer education Safe products and services Precautionary approach Dialogue with public Auditing
The Business Council for Sustainable Development (BCSD)	Declaration (1992)	Open markets Price correction Expanded concept of stakeholders Ecoefficiency
Keidanren	Appeal on Environment (1996)	Environmental ethics Ecoefficiency Voluntary efforts Partnerships with citizens, NGOs, government Global warming Waste reduction Management and auditing

use, and the greater attention to a wider range of stakeholders in defining the roles of corporations.

Here we have two potentially solid points of reference for leveraging change among the quicksands of sustainability-speak. The first is the polluter pays principle, where prices are adjusted to account for the unintended and uncompensated (external) effects of our actions on society, future generations and the environment. How would the price of pesticides be affected if farmers had to pay for the costs of wider damage to health and the environment? And how would these new and presumably higher prices affect the benefits of pesticide use relative to alternative (and currently less profitable) pest control technologies? Secondly, how would a greater involvement of stakeholders other than the traditional groups – shareholders and customers – affect the strategic and technological trajectory of pesticide manufacturers? Four full chapters and a good section of another chapter in this book are devoted to pondering these key questions, so it is fitting that we put these issues in scientific and historical context.

Polluter pays, full-cost accounting, environmental economics

The polluter pays principle is 'one of a small group of environmental principles which have received wide international acceptance' (Baldock and Bennett, 1991). Although industry has opposed every piece of Polluter Pays legislation which attempted to apply green or ecotaxes to products with high external costs, some industry leaders are attracted to the concept as a cornerstone of free market environmentalism, and a more efficient alternative to command-and-control regulation. Ciba's Alex Krauer declared himself to be 'all for the polluter pays principle' (*Der Spiegel*, 2 October 1989). Frank Popoff and David Buzzelli, respectively chairman and vice president of environment, health and safety of Dow Chemical, wrote in 1993 that full-cost accounting 'may be the most important step down the path to sustainable development' and 'one real alternative to today's adversarialism, command and control, overlegislation and overregulation, and national environmental self-interest'. They called for a 'healthy debate among all sectors of society of the options and possibilities that full-cost accounting creates'. The rather unanimous support for full-cost approaches in sustainable development in the BCSD and Rio declarations (see Table 1.3) bodes well for leveraging change.

In agriculture, there is almost universal agreement that cost internalization would be a good thing: farmers should be confronted with the real costs of their actions, whether that is nutrient enrichment and fish declines in the Gulf of Mexico from fertilizer runoff, or bird declines due to pesticide use. This rationale has reached high places; the Council of Economic Advisors recently proposed to the US president that federal policy measures should incorporate environmental and public health values into

Table 1.3 *Business and UN declarations on full-cost accounting and the polluter pays principle*

BCSD Declaration	The prices of goods and services must increasingly recognize and reflect the environmental costs of their production, use, recycling, and disposal. This is fundamental and is best achieved by a synthesis of economic instruments designed to correct distortions and encourage innovation and continuous improvement, regulatory standards to direct performance, and voluntary initiatives by the private sector.
Rio Declaration	*Principle 16.* National authorities should endeavour to promote the internalization of environmental costs and the use of economic instruments, taking into account the approach that the polluter should, in principle, bear the cost of pollution, with due regard to the public interest and without distorting international trade and investment.
Agenda 21	*Chapter 30, article 9.* Governments, business and industry, including transnational corporations, academia and international organizations, should work towards the development and implementation of concepts and methodologies for the internalization of environmental costs into accounting and pricing mechanisms.

farmers' decision-making because: 'When the application of fertilizers and pesticides imposes off-site costs, farmers can only be expected to make efficient decisions if they are themselves confronted with these costs' (*Economic Report of the President*, 1995). Many groups which are fundamentally opposed to pesticides are convinced that, once pesticide prices are adjusted to account for external costs, the relative profitability of pesticide-dependent and alternative farming systems will look drastically different.

The devil, as ever, lies in the details, and a whole host of contentious details have prevented progress on applying the polluter pays principle to pesticide externalities, as they have with other sources of pollution. For figures of total external costs of pesticide use, we are stuck with the rather wild but frequently cited estimates for the US compiled by Pimentel et al (1991, 1993) and recently updated by Steiner et al (1995). The field of environmental economics which has sprung up around free market environmentalism has not yet been used to revisit Pimentel's work. The pesticide industry and its regulators thus have a scanty information base from which to follow through on either the BCSD declaration or Agenda 21, and cynics would argue that it is to industry's advantage to recognize this particular stone but to leave it unturned. They have, after all, become used to working the existing set of regulations for competitive advantage, and see in full-cost accounting the prospect of punitive taxation affecting some of their star products.

It is clear that in the pesticide industry, no systematic attempt has been made to calculate the environmental and social costs of pesticide production, use and disposal. Even the minute fees and taxes on chemical inputs in the US state of Iowa – introduced not as a full polluter pays measure but to fund research and extension on alternatives – was fiercely resisted by agrochemical interests (John 1994). Efforts seem largely directed at rebutting the work of Pimentel from the US and Pingali from the Philippines, rather than coming up with alternative figures. Popoff and Buzzelli's call for a healthy debate among all sectors of society has not begun, even in the companies' own agrochemical divisions and subsidiaries. A discussion with company leaders in 1995 prompted arguments that they 'already have full-cost accounting in place' because national pesticide regulations take products with unfavourable cost-benefit profiles off the market.

Before environmental economics can be a leverage point for more sustainable strategies in the pesticide industry, there is clearly a need for a thorough analysis of the theory and application of full-price accounting when applied to pesticides. These are the objectives of Chapters 3 and 4 by David Pearce and his group at University College London.

Stakeholder analysis and involvement

Another point of broad agreement between Agenda 21 and industry declarations on sustainable development is the need for a more inclusive and more participatory approach to the way companies build their strategies and select their technologies (see Table 1.4). In the main work of *Changing Course*, Schmidheiny and the BCSD (1992, 86) came out even more adamantly in support of stronger stakeholder relationships:

> *Considering stakeholder involvement to be legitimate and strategically important requires more effort than traditional public relations or information-sharing responses. New forms of collaboration are needed, including focus groups, advisory panels, forums for dialogue, and joint ventures. Building stakeholder involvement in the context of sustainable development extends the idea of corporate responsibility in time and space. Companies now have to consider the effects of their actions on future*

Table 1.4 *Business and UN declarations on stakeholder involvement in the management of industry and technology*

BCSD Declaration	We must expand our concept of those who have a stake in our operations to include not only employees and shareholders but also suppliers, customers, neighbours, citizens' groups, and others. Appropriate communication will help us to refine continually our visions, strategies and actions.
Keidanren	*Charter for Good Corporate Behaviour,* 3. Corporations will communicate not only with shareholders but also with society as a whole, actively and fairly disclosing corporate information.
Agenda 21	*Chapter 30, article 26.* Business and industry should increase self-regulation, guided by appropriate codes, charters and initiatives integrated into all elements of business planning and decision-making, and fostering openness and dialogue with employees and the public. *Chapter 31, article 1.* The cooperative relationship existing between the scientific and technological community and the general public should be extended and deepened into a full partnership. The public should be assisted in communicating their sentiments to the scientific and technological community concerning how science and technology might be better managed to affect their lives in a beneficial way.

generations and on people in other parts of the world. Prosperous companies in a sustainable world will be those that are better than their competitors at 'adding value' for all their stakeholders, not just for customers and investors.

Social investor groups are also looking at stakeholder involvement and participation as a cornerstone of global corporate responsibility:

> *An increasingly globalized economy requires a redefinition of the concept of stakeholder. The Principles make the community, rather than the corporation, the starting point in this new definition of stakeholder. In order for communities to be sustainable, all members of the community need to be recognized as stakeholders in the community as a whole. Corporations are stakeholders in the community along with consumers, employees, stockholders, and the community at large.*
>
> Global Corporate Accountability Issue Group of the ICCR

These statements appear to be very much at odds with perceptions of the chemical industry, which is seen to have a lamentable record of public dialogue. The industry's attempts at communication have been typically thinly disguised attempts at marketing to recruit people to the industry's world view rather than to engage in a meaningful discourse. The residues of the PR mindset are seen in the BCSD's call for appropriate communication.

The pesticide industry in particular has historically taken an extraordinarily narrow view of who is a genuine stakeholder in setting the pesticide agenda, dividing the world into those with them and those against them. Although it was not that long ago when chemical companies were the outsiders of agriculture, we can frequently read statements that pesticide manufacturers and dealers are the insiders of agriculture, and that environmental groups and other outsiders in the non-farming public have undue influence on pesticide policy. Neil Strong of Ciba stated in 1994: 'We must prevent those outside of production agriculture from making changes [to the 1995 Farm Bill] that will surely negatively affect the best food production system in the world.' DowElanco added in the same year:

> *Many environmental and advocacy groups have become specialists in affecting public policy... These well-established groups may have limited agricultural experience. Yet, they have become increasingly involved in agriculture policy making... It is time for all who have a stake in agriculture to ensure they are heard...*

That non-agribusiness stakeholders should want a say in how their food is produced and the land is managed is considered by agribusiness to be irresponsible, not only because they are not 'in' farming, but because they are not qualified scientists. The elevation of science to the position of grand arbiter has a long tradition in the pesticide debate and continues today.

Having experienced and suffered from a loss of social consensus with its pesticide technologies in the 1960s and 1970s, there were signs that companies would make a far stronger effort to ensure that their newest technological focus – genetically engineered seeds – would evolve in parallel with public and institutional change. Ciba-Geigy's 1990 *Annual Report* stated that 'Broadly based public discussion [about biotechnology] will be welcome, because it is the only way to arrive at a lasting social consensus on the vital research questions.' But by 1996, the strong-arm tactics of Ciba and Monsanto to get genetically engineered maize and soybeans accepted in the EU (in the name of greater agricultural sustainability) produced a predictable societal backlash. Things were so bad that one journalist commented: 'Companies such as Monsanto and Ciba-Geigy have begun to act as though people are a peculiar irritant to the achievement of their aims of making money from selling biological [sic] products to farmers.' (Charles Arthur in *The Independent*, 21 October 1996).

Clearly, very little has been learned from the three decades of the pesticide era in which the same 'solutions' were pushed over and over again. In a statement of extraordinary arrogance, an executive from a US seed company was quoted as saying that 'we can't let special interest groups set the agenda for what we'll be allowed to do with science and technology that make improvements for mankind' (*Progressive Farmer*, October 1996, p11). But are these companies not right to be wary of opening a Pandora's box of stakeholder demands on their organizations, considering that the industry and its products are consistently distrusted by the majority of the general public? If companies are intent to defend the status quo at all costs, that would be correct. But as we have seen with the response in Europe to crop biotechnology, those costs can be considerable in terms of perpetuating the pesticide industry's image of putting profit ahead of caution and the purity of the food supply.

A deeper understanding of the dialogue around the pesticide and sustainability issue which involves all stakeholders would allow, for once, a balance of interests and therefore a balance of priorities in choosing how to develop new technologies. It could point towards innovations which support agricultural sustainability with a much more solid social consensus than that which prevails for the current basket of chemical and genetically engineered technologies. It would also confront stakeholders who are critical of pesticides with some uncomfortable realities, which may be much closer to home than the corporate boardroom. Stakeholder dialogue cuts both ways. Involving the entire system of stakeholder interests means that we can no longer blame outside circumstances for our problems. As Senge (1990, p67) writes: 'Systems thinking shows us that there is no outside; that you and the cause of your problems are part of a single system.' All players along the agrifood chain may have to come face to face with issues of sustainable consumption, such as the 'demand' for the year-round availability of cosmetically perfect fruit and vegetables.

> *Take, for example, agriculture, which, in the final analysis, produces surpluses of meat, grain, potatoes as well as other surpluses with the aid of chemicals. This is economic nonsense. The question should be posed as to whether we really have to eat fresh strawberries in the middle of winter. But the solution to these problems does not lie with the chemical industry. A 'pact of reason' is necessary. What is required is a new solidarity between industry, the community and public officials. Each and every one must bear part of the responsibility. Together we must develop perspectives as to what we mean when we talk of a better future. This is the only way.*
>
> Alex Krauer, CEO of Ciba-Geigy's worldwide operations
> (*Der Spiegel*, 2 October 1989)

Two approaches to exploring the stakeholder dialogue for leverage points are taken in this book. In Chapter 6, Mayhew and Alessi apply rigorous qualitative methodology derived from family therapy in order to understand the dysfunctional nature of the pesticide debate. In Chapter 7, Morley and Franklin report on a Future Search workshop held in 1996, in which a diverse group of stakeholders 'backcast' from an agreed ideal future to plan the necessary actions of industry, government and other stakeholders for creating that future.

Summary: An Interdisciplinary Approach

> *Modernity is the transition from fate to choice. But at the same time it dissolves the commitments and loyalties that once lay behind our choices. Technical reason has made us masters of matching means to ends. But it has left us inarticulate as to why we should choose one end rather than another.*
>
> Jonathan Sacks, Chief Rabbi of Great Britain (Sacks, 1991, p32)

This book is a case study of an industry which finds itself with a vision of sustainable development and an entire product range which seems to be an anathema to that vision – pesticides – and a uniform perspective of its role in agriculture. We hope that this study is the beginning of an articulation of the choices that face the leaders, regulators and other stakeholders of this particularly controversial industry if they are to choose a course of sustainability.

The study is about analysing the implications of what industry has already said on the subject of sustainable development. From industry's statements and declarations, we have identified potential leverage points in the highly polarized pesticides and sustainability debate in the form of full-cost accounting and stakeholder participation. But unless these topics are understood in a rigorous way and explored for their potential to reframe the problem for industry and policy-makers, there is every chance that corporate statements about sustainability will be left on the shelf. Accusations of corporate greenwash – the immune response of the status quo – would then be justified.

New options for this industry and its regulators which work in the long term will require a holistic understanding of the system – exploring the *context* of pesticide production in the greening of industry, and the context of pesticide use in sustainable agriculture, around concepts and vocabulary understood by all sides. Only then can we develop options and solutions which are sustainable or, in the words of Wendell Berry, 'solve for pattern': that is, solve for more than one problem while creating no new ones. Understanding the forces at play in systems, furthermore, means crossing a number of disparate disciplines. Paul Hawken writes in the foreword to David Wann's ground-breaking *Deep Design* (1996 pp xi–xii) that 'we cannot practice architecture without knowledge of forestry and energy issues, that chemical engineering without epidemiology and biology is inexact and lacking, that transportation systems that do not take into account community, family and climate are not systems at all'. Likewise, we cannot talk about the pesticide industry without developing literacy in agriculture, ecology, environmental philosophy, commerce, economics, and sociology.

Holism and interdisciplinarity are often considered symbols of poor-quality, value-laden, woolly science. Multiple perspectives can also be misused to question all values and encourage moral relativism (Linstone, 1994). But without these different perspectives, our ability to innovate for sustainability is hindered by imbalance or downright ignorance. We become the prisoners of a single perspective. So often we see both industry and its regulators talking environment with only the language of toxicology, talking agriculture with only the language of crop protection, and talking sustainability with only the language of economics. The challenges are trivialized as simply technological problems, and the road to green business becomes the stuff of bulleted handbooks with subheadings such as 'The Five Steps of Corporate Environmental Leadership'. But as Aldo Leopold wrote a half century ago: 'No important change in ethics was ever accomplished without an internal change in our intellectual emphasis, loyalties, affections, and convictions' (Leopold, 1949, pp 209–210). Let's see what the chapters that follow do to our intellectual emphasis, loyalties, affections and convictions.

References

Athanasiou T (1996) 'The age of greenwashing'. CNS 7(1), pp 1–36.

Avery DT (1995) *Saving the Planet with Pesticides and Plastic*. Hudson Institute, Indianapolis Indiana.

Baldock D and Bennett G (1991) *Agriculture and the Polluter Pays Principle: a study of six EC countries*. Institute for European Environmental Policy, London, UK.

Beus CE and Dunlap RE (1991) 'Measuring adherence to alternative vs conventional agricultural paradigms: a proposed scale'. *Rural Sociology* 56(3) pp 432–460.

Bloom DE (1995) 'International public opinion on the environment'. *Science* 269, pp 354–358.

Bruno K (1992) *The Greenpeace Book of Greenwash*. Greenpeace Communications Ltd, London.

Buttel FH (1992) 'Environmentalism: origins, processes, and implications for rural social change'. *Rural Sociology* 57(1) pp 1–27.

Ciba-Geigy (1993) 'Sustainable agriculture: we know it's important. What does it mean?' *Ciba Broadcast* July 1993, Ciba-Geigy Corp, Greensboro NC.

Constanza R (1988) 'Social traps and environmental policy'. *Bioscience* 37, pp 407–412.

Deal C (1993) *The Greenpeace Guide to Anti-Environmental Organizations*. Odonian Press, Berkeley CA.

DowElanco (1994) *The Bottom Line – 1994 Issue on Sustainable Agriculture*. DowElanco, Indianapolis IN.

Easterbrook G (1995) *A Moment on the Earth: The Coming of Age of Environmental Optimism*. Viking, New York.

Edwards D (1995) *Free To Be Human: intellectual self-defence in an age of illusions.* Resurgence, Green Books Ltd, Dartington, UK.

Hansen JW (1996) 'Is agricultural sustainability a useful concept?' *Agricultural Systems* 50, pp 117–143.

Hawken P (1993) *The Ecology of Commerce: a declaration of sustainability.* HarperBusiness, NY.

Hill SB and MacRae RJ (1995) 'Conceptual framework for the transition from conventional to sustainable agriculture'. *Journal of Sustainable Agriculture* 7(1), pp 81–87.

John D (1994) *Civic Environmentalism: alternatives to regulation in states and communities.* CQ Press, Washington DC.

Johnson G and Scholes K (1993) *Exploring Corporate Strategy* (third edition). Prentice-Hall, Hemel Hempstead, UK.

Krauer A (1992) 'Sustainable development – a window of opportunity for progressive companies'. *Ciba-Geigy Journal* 2, pp 6–11.

Leopold A (1949) *A Sand County Almanac and Sketches Here and There.* Oxford University Press, NY.

Linstone HA (1994) 'New era – new challenge'. *Technology Forecasting and Social Change* 47, pp 1–20.

MacRae RJ, Henning J and Hill SB (1993) 'Strategies to overcome barriers to the development of sustainable agriculture in Canada: the role of agribusiness'. *Journal of Agricultural and Environmental Ethics* 6, pp 21–50.

Magretta J (1997) 'Growth through sustainability: an interview with Monsanto's CEO, Robert B. Shapiro'. *Harvard Business Review* January–February 1997, pp 79–88.

Pearce F (1992) 'How green are the multinationals?' *New Scientist*, 23 May 92, p43.

Pimentel D, McLaughlin L, Zepp A, Lakitan B, Kraus T, Kleinman P, Vancini F, Roach WJ, Graap E, Keeton WS and Selig G (1991) 'Environmental and economic effects of reducing pesticide use'. *BioScience* 41(6), pp 402–9.

Pimentel D, Acquay H, Biltonen M, Rice P, Silva M, Nelson J, Lipner V, Giodano S, Horowitz A and D'Amore, M (1993) 'Assessment of environmental and economic impacts of pesticide use'. In *The pesticide question: environment, economics, and ethics*, eds D Pimentel and H Lehman, pp 47–84. Chapman & Hall, New York.

Popoff F and Buzzelli DT (1993) 'Full-cost accounting' *Chemical and Engineering News* 11 January 1993, pp 8–10.

Saunders T and McGovern L (1993) *The Bottom Line of Green is Black: strategies for creating environmentally sound businesses.* HarperCollins, New York.

Sacks J (1991) *The Persistence of Faith: religion, morality and society in a secular age.* Weidenfeld and Nicolson, London (1990 BBC Reith Lectures).

Schmidheiny S with the Business Council for Sustainable Development (1992) *Changing Course: a global perspective on development and the environment.* MIT Press, Cambridge MA.

Senge PM (1990) *The Fifth Discipline: The Art and Practice of the Learning Organization.* Currency Doubleday, NY.

Stauber J and Rampton S (1995) *Toxic Sludge is Good for You! Lies, Damn Lies and the Public Relations Industry.* Common Courage Press, Monroe, Maine.

Steiner RA, McLaughlin L, Faeth P and Janke RR (1995) 'Incorporating externality costs into productivity measures: a case study using US agriculture'. In V Barnett et al (eds) *Agricultural Sustainability: economic, environmental and statistical considerations.* John Wiley & Sons, Chichester, UK, pp 209–230.

Strong N (1994) 'Speaking out for agriculture'. *Ag Retailer*, December 1994, p8.

Thompson K (1993) 'The Rio Declaration on Environment and Development'. In M Grubb et al (eds) *The Earth Summit Agreements: A guide and assessment.* Earthscan, London, pp 85–95.

Thompson PB (1995) *The Spirit of the Soil: agriculture and environmental ethics*. Routledge, London.

Tombs S (1993) 'The chemical industry and environmental issues'. In *Business and the Environment: Implications of the New Environmentalism* (ed) D Smith, pp 131–149. Paul Chapman, London.

Turner RK, Pearce D and Bateman I (1994) *Environmental Economics: an elementary introduction*. Harvester Wheatsheaf, Hemel Hempstead, UK.

Walley N and Whitehead B (1994) 'It's not easy being green'. *Harvard Business Review* May–June 1994, pp 46–51.

Wann D (1996) *Deep Design: Pathways to a Liveable Future*. Island Press, Washington DC.

Willers B (1994) 'Sustainable development: a new world deception'. *Conservation Biology* 8(4), pp 1146–8.

2

Pesticides in World Agriculture: Causes, Consequences and Alternative Courses

Jules Pretty, William Vorley and Dennis Keeney

He who only knows his side of the case, knows little of that.

John Stuart Mill, *On Liberty*, 1859, p2

Every truth has two sides; it is well to look at both, before we commit ourselves to either.

Aesop, 'The Mule', *Fables* (6th century BC?) tr. Thomas James

Introduction

Technologies can follow a number of different paths, depending on the assumptions, interests and world views of those who develop and support them (Clark and Lowe, 1992; Tenner, 1996, p274; Vorley and Keeney, 1995). To understand the path which pesticides have taken in agriculture, we must understand the assumptions on which their developers – the scientists in private and public institutions – have based their work. And to challenge the sustainability of that path, we must challenge those assumptions and metaphors carefully, and try to avoid substituting one stream of naïveté with another (Linstone, 1994).

It is not difficult to see what drives a pesticide company to proclaim their contributions to sustainable development. Populations are growing and aspirations are rising. New mouths have to be fed and urbanizing populations have to be sustained with food and clothing. Pests compete with humans for this extra food and fibre. So by focusing our technology on controlling these menacing insects, diseases, weeds and other scourges, we can keep the upper hand in the pest war. In fact, the demographic challenge is the first and last word of many justifications for the production, sale and use of pesticides. Others include the quest for improved farm productivity, which reduces the price of food and thereby supports competitive agricultural exports, freeing up economic activity for consumption in other parts of the economy, and land which would otherwise have been farmed for wildlife and nature.

To get a flavour of the agribusiness world view, here are some daily messages from a 1996 desktop calendar produced by the US Agricultural Retailers Association's 'Food for Thought' campaign (US ARA, 1996):

> *The earth's population is expected to double to 11 billion by the year 2050, which will require that each farmer be twice as productive.*
>
> *In order to survive, crops must compete successfully with 30,000 weeds, 10,000 insect species and 3000 worm species, along with rodents, wildlife and a variety of diseases and fungi.*
>
> *Modern agricultural products contribute to an affluent society. Since most families do not produce their own food, we have a greater opportunity to pursue leisure or alternative activities.*
>
> *Americans enjoy the safest, most wholesome food supply in the world.*

This justification – of keeping the pest menace at bay to put food in mouths and money in pockets – explains the pesticide manufacturers' sense of mission as supporters rather than enemies of agricultural sustainability. If this were the only interpretation of the pesticides–sustainability debate, there would be little redesigning of the industry necessary for sustainable development. Companies could continue their focus on reducing the impact of conventional practices and replacing environmentally damaging products with more benign technologies. This is more of the same, just with greater benefits and fewer risks. Stuart Hill and Rod MacRae call this 'efficiency' or 'substitution' initiatives (MacRae et al, 1993; Hill and MacRae, 1995). But readers of this book will be aware that there are fundamental disagreements in society about whether pesticides are or are not a justified response to the challenge of sustainable agriculture. The Food for Thought calendar reflects the world view of just one of the actors in the pesticides and sustainability debate, though – as the developers and promoters of technology – this is a very influential world view.

It is tempting to attribute the different outlooks on pesticides to different moral positions. We could label agribusiness as anthropocentric productionists, and label environmentalists as biocentric agrarians. But this would be to miss the point, that world views are not necessarily derived from core beliefs but are a consequence of exposure to only one set of beliefs. World views of organizations – from chemical companies to environmental groups – are self-reinforcing in that research is selectively funded, information is selectively acquired, and personnel are selectively hired in ways which reinforce the dominant paradigm.

Our purpose in this chapter is, therefore, to present a range of alternative views on the *drivers* of pesticide use, the *consequences* of pesticide use, and the potential to significantly *reduce* pesticide use in farming in the context of sustainable agriculture. These aspects are strongly interrelated. If farmers depend on pesticides to protect their crops from the ravages of pests and provide the world with affordable and plentiful food, then our perception and tolerance of associated risks will be much more different than if pesticides were just one technological option for pest control, or a crutch to prop up other inherently unsustainable agricultural practices. We seek to question the assumptions that underlie the prevailing rationale for pesticide production and use in both industrialized or developing and agricultural economies.

Drivers of Pesticide Use

The majority of the two million tonnes of pesticides produced in the world are applied to crops in OECD countries. One third of global expenditure and a quarter of the tonnage is accounted for by the giant US pesticide market. Multinational pesticide companies cannot survive without good penetration into the markets of the US, EU and Japan. In developing countries, the majority of pesticides are consumed in the cash grain export sectors of Brazil and Argentina, or in smallholder cotton production in India and China. Few pesticides – by global standards – are used to protect domestic food crops in developing countries.

Farmers are not fans of pesticides. Nearly two out of every three Iowa farmers, for example, believe that agriculture depends too much on fertilizers and pesticides (Iowa State University, 1994). Many farmers are also aware that few of them have benefited financially from new technologies over the long term. But, as Buttel (1993) writes:

> *A good number of [farmers] have chosen to go to considerable lengths to preserve their prerogative to use agricultural chemicals. Any social scientist or policy maker who wishes to understand the social and ethical issues involved in reducing chemical usage must therefore confront the fundamental question of why, especially from World War II to the present, farmers have tended voluntarily to use plant protection chemicals.*

Confronting the structural causes of this entrenched position of pesticides in agriculture requires reflection on the operating principles of the food and agricultural system (Allen and Sachs, 1993). We present in this section alternative views of those operating principles, which represent a diversity of political and economic standpoints. The farmer may be portrayed as a victim, and agribusiness as aggressors, or both may be seen as entirely rational economic actors responding to changing consumer demands and global economic trends. The common thread, which is absent from most rhetoric on both sides of the pesticide debate, is the synergism between pesticides and a raft of economic, technological, political and commercial movements which have set agriculture into a pattern of chemical dependence. Pesticides may be the *cause* of some aspects of agriculture's unsustainability, but they are more likely to be a *consequence* of forces on agriculture which are generating a whole range of social and environmental outcomes, which themselves are seen by many observers to be inherently unsustainable. We start with the most fundamental trend in agriculture – the marginalization of farming – which has forced the specialization and simplification of farming systems.

The marginalization and specialization of farming

Forces affecting agriculture throughout much of the 20th century, in which profit margins in farming have been steadily shrinking, leave farmers little choice but to move fast in the technological treadmill as the price of survival. Buttel (1994) describes the context of this process succinctly:

> *After World War II, the nature of agricultural production began to parallel that of the manufacturing industry. Agricultural production came to be based on industrial inputs (eg chemicals, machinery) produced within the dominant industrial complex. Farm production became specialized at both the enterprise and regional levels, and took the form of mass production of undifferentiated commodities. Particular types of agricultural technology – especially chemical inputs and crop varieties bred to be*

responsive to these inputs – became rapidly generalized across the industrialized world (and, through the Green Revolution, much of the Third World as well). Agricultural productivity increased significantly through a treadmill of technological change that increased the capital intensity of agriculture.

The 'technological treadmill' (Cochrane, 1958) is particularly a problem in markets with relatively inelastic supplies, such as agriculture, and led to agricultural surpluses in the US long before the pesticide era. The benefits of pesticides were immediately recognizable to farmers, especially as genetic improvements of crops increased the potential productivity of yield-protecting technologies. Together with fertilizers, pesticides allowed farmers to break away from crop rotations and diversified crop and livestock farms. Until the mid 1950s, ley farming (grass and arable crop rotation) had been the basis of the traditional mixed crop and livestock farming in temperate agriculture. The main advantage of the ley system was to keep weeds, diseases and insect pests of the arable crop in check, and to build soil fertility (Tovy, 1990, p228). Pesticides allowed the simplification of the agroecosystem, especially the replacement of considerable acres of ley farming with continuous cultivation, so that farmers could realize the economies of scale from specialized crop or livestock production. However, in a technological treadmill, profits accrue only to the early adopters. As more farmers adopt a given technology or method, production increases and prices fall, first squeezing and then eliminating the previous margins of profit. The incentive for later adopters is survival rather than profitability, and those who adopt too late do not survive. For farmers, the long-run effect has been increased operating costs due to pesticide purchases and a heavier regulatory burden, while profits were bid away over time as more farmers adopted the new technology and markets adjusted to greater output. Higher farm incomes are just capitalized in the price of land.

In the process, many agricultural activities previously undertaken on the farm have relocated to other parts of the food system. The shift of agriculture from farm to non-farm reduces returns that cover opportunity costs and requires farmers to either increase production or utilize their excess management and labour in non-farm pursuits (Goodman and Redclift, 1991, pxvii; Smith, 1992). Increasing production to sustain a full-time farmer and family necessitated larger farms. The overwhelming majority of farms that once existed in the US no longer exist and production is highly concentrated among the remaining farms. Between 1890 and 1990, the percentage of the US population living on farms fell from 40 per cent to less than 2 per cent and farm labour now accounts for only 2.8 per cent of the labour force. These small numbers have not stabilized yet: the farm resident population declined by 24 per cent during the decade of the 1980s (Unnevehr, 1993). Those living on farms today earn more than half their income from non-farm sources. The largest 3 per cent of farms in the US now produce half of the value of total production. Similar trends have occurred in Europe and Japan. In Japan the total number of farm households has declined from 6.2 million in 1950 to less than 3.5 million in 1995, of which only 16 per cent are full time. In the UK, about 144,000 full-time holdings produce more than 97 per cent of the total output.

We should note that the increase in farm size (and consequent loss of family farmers) has no longer much to do with improving efficiency:

Thus farm sizes have grown not to achieve greater efficiency, but primarily because expansion has been the only way to maintain their incomes, giving rising costs that relentlessly shrink profits per acre. Moreover, those farmers who can, will expand to continue increasing their incomes... As a result of these pressures, over half of the agricultural commodities produced in America are now grown on farms larger than can be justified on grounds of efficiency.

(Wessel, 1983, p50)

Whether we approve of these trends or not – and there are strong arguments that sustainable agriculture should sustain farmers as well as natural resources (see Comstock, 1987) – there is no doubt that pesticide dependence usually increases as farm size increases and management systems become simplified.

> *Large farms – the type which American farms are tending to become – are thus potentially lucrative, but perhaps also highly vulnerable enterprises. Farmers may be able to guard themselves against disastrous price declines through political action leading to government price support programmes. Their vulnerability to loss of income from price declines is mirrored, however, by a similar sensitivity to any factor that decreases yields, such as insects. The trends of American agriculture that led to larger farms may thus be creating businesses particularly sensitive to damage to insect pests.*
>
> *From this vantage point, we can begin to see through the fog of rhetoric surrounding insect control technologies: the purpose of insect control technology is only secondarily to protect crops from damage; its primary purpose is to protect farm firms, especially large ones, from bankruptcy.*
>
> (Perkins, 1982, pp 231–2)

More than half of US farmland is operated by persons within ten years of retirement. This itself has implications for chemical use: a 1989 survey of 2000 Iowa farmers showed that farmers younger than 40 were more likely to experiment with chemical-saving methods than those of 60 or more years; older farmers were more likely to use only two of 11 chemical-saving techniques (*Des Moines Register* 4 December 1990). In 1995, 43 per cent of Japanese farmers were over 65 (Japan-MAFF 1995).

To summarize, pesticides have been a causal factor and also a farmer's response to shifts in agricultural activity off the farm. Pesticides, especially herbicides, have contributed to a massive 'release' of labour from agriculture since their widespread adoption in the 1960s and 1970s; post-war labour shortages have also been a factor in the rapid uptake of this technology. Furthermore, scarcity of skilled labour is now viewed by farmers as a barrier to the adoption of alternative, more sustainable farming practices (Pfeffer, 1992; Lighthall, 1996). For the remaining 'efficient' farmers, it has become difficult to spread management over more and more acres without drastically simplifying the farm enterprise and minimizing risk. An alternative response to input-dependent farm expansion – intensive use of biological and management resources on smaller diverse farms – also has worked for a few very skilled and highly motivated individuals. But these farmers have had little support from the technology-driven juggernaut of university research, extension and agribusiness.

Intensive livestock production and industrial uses of crops

The maize and soybean submarkets in the United States are the richest sectors of the world's biggest pesticide market. American farmers spent $3.2 billion on pesticides for these two crops, accounting for 44 per cent of all US crop protection sales (*Agrow* 257, pp 16–17) and approximately 12 per cent of worldwide sales.

The linkage between pesticide use and livestock is rarely discussed. But it should be noted that maize and soybeans are not primarily food crops but *feed* crops (see Table 2.1). Two-thirds of the 1994–95 US maize crop was used for animal feed, including ruminant cattle. A further 5 per cent of the US maize harvest – 11 million tonnes – was used for high-fructose corn syrup (HFCS), most of which sweetens colas and other drinks. Americans drank about 53 billion litres of carbonated soft drinks in 1994 which included 6.5 million tonnes of HFCS. Six per cent of maize was converted to alcohol to blend in fuel for gasoline-powered vehicles, while only 1 per cent was used for cereal and other food products (USDA-ERS, 1994). Of the 17 per cent of the crop exported,

Table 2.1 *US pesticide use by feed, food and fibre groups*

Crop group	*Total pesticides used, 1992 (metric tons)*
Feed (corn, sorghum, soybeans, wheat)	159,000
Food (rice, peanuts, potatoes, other vegetables, citrus, apples)	75,000
Fibre (cotton)	26,000

Source: Porterfield et al, (1995)

much is destined for intensive animal production in Japan, Korea, Taiwan and China. Of the total world trade in coarse grains (mainly maize), 70 per cent is for animal feed rather than direct human consumption (Watkins, cited in PANOS, 1996).

The cash grain-livestock system has had major environmental consequences (Lockeretz, 1994). It represents the mining of nutrients in one area to feed a surplus of nutrients in another. Declining soil quality may require more pesticides to compensate for plant health problems. Furthermore, the production of export feed grains on huge farms in countries such as Brazil has displaced countless peasant farmers, pushing land hungry farmers into ecologically fragile margins, such as the Amazon region. Brazil and Argentina will plant around 19 million hectares of soybeans in the 1997/8 season; the majority of the harvest is exported to Europe (Lang, in press).

The list of negative impacts of the current feed grain-livestock system does not stop there. At the grain production end are problems of unsustainable water consumption, soil erosion and salination, fertilizer use and runoff, which in many areas of the world – including the Gulf of Mexico – contributes to coastal eutrophication and oxygen-deprived 'dead zones' which cannot support marine life. Soil erosion can be reduced considerably by minimum tillage systems (which leave the soil less disturbed and with a greater covering of crop residue), but these systems are *more* reliant on herbicides for weed control, and encourage raising grain crops on land which would otherwise be better suited to grass-based agriculture. At the livestock end is a reliance on pharmaceuticals, including antibiotics given as subtherapeutic feed supplements, which in turn hastens drug resistance in human pathogens. Many commentators also decry the lack of animal welfare in intensive confinement systems.

These signs of unsustainability cannot be blamed on pesticides. Our point is that pesticides have been part of a suite of technological and commercial developments which have allowed the system to develop and which keeps the system running. The massive use of pesticides in cash grain farming has much more to do with 'feeding the feedlot' than with 'feeding the world'.

Production subsidies

Countries have had good reasons to subsidize agriculture and to shelter farmers from the drastic restructuring which technological innovation would force on farming in an unregulated market. These reasons include maintenance of rural economies and culture, appeasement of powerful political interests, drives for national self-sufficiency, or stabilization of agricultural markets. Protection and subsidies compensate farmers for the tendency of technology-driven gains in productivity to drive down prices and produce huge agricultural surpluses. Given the rate of technological advance, this protection has proved hugely expensive.

The total support in the OECD is some $180 billion, which represents about $16,000 per full-time farmer-equivalent, or $179 per hectare of farmland. There is, though, great variation between countries: from the high level of support in Japan of 71 per cent of the total value of agricultural production to just 3 per cent in New Zealand (OECD, 1993). The massive levels of support given to producers, particularly in OECD countries, in the form of support prices and commodity programmes, import controls, international surplus disposal and domestic food-welfare programmes have been one of the leading stimuli of pesticide use. A classic example of the impacts of subsidies on pesticide use is in rice (Anderson and Blackhurst 1992, p162). Japanese taxpayers and consumers pay $13 billion each year for their rice farmers to receive a price about ten times above the world price for their crop, and farmers in Taiwan and Korea also receive considerable support. This leads to an incredible statistic. In 1990, 53 per cent of the value in global rice insecticides was applied to only 2 per cent of the world rice crop – in Japan, Korea and Taiwan. The high price which north-eastern Asian farmers can expect for their crop makes it very profitable to apply yield-increasing and yield-protecting technologies, even when they contribute very little to the level of production. Thus even if rice farmers in Japan and the Philippines have the same attainable yields of six tonnes per hectare, the Japanese farmer only needs a 0.3 per cent yield benefit to break even on a $50 pesticide application, while the Filipino must get a 5 per cent yield benefit.

Production subsidies clearly shift farmer behaviour towards high levels of preventative 'insurance' usage of pesticides (see Table 2.2) – they are recognized as a serious disincentive to integrated pest management (IPM).

Production subsidies not only have negative influences on the use of chemical inputs. The European Union's Common Agricultural Policy (CAP), providing high support prices (amounting to over $45 billion in 1997), led to massive surpluses that were exported only with additional subsidies. The 'dumping' of these surpluses from the EU – along with the subsidized exports from the US under the Export Enhancement Programme – have had catastrophic effects on developing country food producers by depressing world prices and masking true market signals.

Another policy tool used to maintain farm incomes and reduce production surpluses is temporary set-aside programmes or more permanent conservation easements which reduce the area of farmed land. Farmers respond by applying more capital inputs such as fertilizers and pesticides to the remaining farm land. Net transfers *away* from agriculture can be as much a threat to the sustainability of agriculture as production subsidies. The taxation of agriculture to provide urban populations with cheap food is especially prevalent in Africa and is linked with overextensification, nutri-

Table 2.2 *Agricultural subsidies and intensity of pesticide use*

Country/Bloc	*Agricultural subsidy equivalent (aggregate producer support-1989)*	*Kilogrammes pesticide applied per farm hectare (1995)*
Japan	71.0 %	17.7
EU–12	30.1 %	4.5
US	18.5 %	2.4
Brazil (1987)	(8.1 %)	0.8

Source: Data compiled from USDA–ERS and Oskam (1995)

ent mining and soil erosion, which drives farmers into fragile marginal lands. Robert Paarlberg (1996) observes that subsidizing farmers is likely to cause environmental damage *off* the farm at the expense of non-farmers, while de facto taxation of farmers causes environmental damage on the farm at the expense of farmers themselves. Subsidies of another kind – of pesticides themselves – were also a factor in serious pesticide overuse in the Green Revolution era in countries such as Bangladesh and Indonesia, but this practice is disappearing under donor pressure and trade liberalization (see Farah, 1994, Ange, 1996 and Chapter 3 of this book). So, again, we see pesticide use as a response to wider economic and political forces which are shaping modern agriculture.

Demand for cosmetic quality

Concentration of buying power in the hands of a few supermarket retailers (see Chapter 5) means large orders for uniform top-quality produce, especially fruits and vegetables. Producers have to meet high cosmetic standards as well as low risk of pesticide residues, in line with perceived consumers' demands. Despite the contradictions inherent in these market signals, there is no doubt that large quantities of pesticides are used to protect cosmetic appearance rather than other aspects of food quality such as flavour and nutritional value. For instance, in apples quality benefits accounted for about one third of total pesticide use in the crop (Babcock et al, 1988). A survey of US food retailers by the US-based Public Voice for Food and Health Policy (Rosenblum, 1994) revealed extremely strict standards of cosmetic quality. One quarter of the surveyed retailers required apples with zero blemishes, while an additional 41 per cent would accept produce only with less than 1 per cent of blemishes. The study also cites a report which concluded that 60 to 80 per cent of insecticide treatments on Californian oranges are used to reduce cosmetic damage on fruit. Fruits and vegetables are the largest consumers of pesticides worldwide, accounting for a quarter of pesticide usage in 1995.

Out of season production of strawberries is an extreme case of pesticide use to preserve quality during storage and shipment. Each hectare of the Florida strawberry crop receives on average around 2.1 kilogrammes herbicides, 21 kilogrammes of insecticides, 150 kilogrammes of fungicides and 470 kilogrammes of soil fumigants per year. In many tropical countries, there is a growing demand for vegetables, especially exotic Northern vegetables such as cabbage, again with high demand for cosmetic quality. Some of the most heavily sprayed regions of the world are the highland vegetable-growing regions of Indonesia, the Philippines and Thailand. Pesticide expenditures may reach over 40 per cent of total production costs. Pesticide applications may continue even after harvest, when the vegetables are loaded on the truck. It is common for these farmers to set aside a parcel of land to grow their own low-input vegetables. For export crops, growers have to be mindful of both standards for pesticide residues and cosmetic quality. A survey of small-scale snow pea growers in the highlands of Guatemala showed that most growers lived in fear that their produce would be rejected for low aesthetic quality. 'The fear is logical', comments the author: 'In 1993, on average, 16 of every one hundred pounds of Guatemalan snow peas produced were rejected due to blemishes' (Thrupp, 1995, p54).

Free trade and global sourcing

Increases in productivity from a shift to intensive farming systems such as large-scale specialized grain production, in which fertilizers and pesticides have allowed a break from dependence on crop rotations and livestock, obviously make a country's agricultural exports more competitive. In a free market, countries where these technologies have not been employed can either go the same way (sometimes described as a 'race to the bottom' or 'the bad driving out the good') or watch their own farm sector decline.

Back in 1940, Sir Albert Howard (1940, p220) was already linking globalized agriculture and domestic pesticide use:

> *The flooding of the English market with cheap food, grown anywhere anyhow, forced the farmers of this country to throw to the winds the old and well-tried principles of mixed farming, and to save themselves from bankruptcy by reducing the cost of production. But this temporary salvation was paid for by loss of fertility. Mother earth has recorded her disapproval by the steady growth of disease in crops, animals and mankind. The spraying machine was called in to protect the plant; vaccines and serums the animal... This policy is failing before our eyes.*

Howard would not have approved of the path chosen to protect European and northeast Asian farming against competition from 'cheap food, grown anywhere anyhow' – price support – which had the effect of stimulating pesticide use. But countries which are now trying to restructure their agricultures along ecological lines, with incentives for reduced dependence on chemical inputs, are facing the prospect of World Trade Organization (WTO) action against trade barriers. Meanwhile, horticultural production is shifting to where labour costs and climate allow lowest cost production and year-round availability on supermarket shelves. This has consequences for pesticide use, worker protection, externalities of long-distance transport, and suppression of local (seasonal) production (Balthasar, 1995; Wright, 1990).

Social pressure

Clean fields have been a sign of a good, hard-working farmer since the earliest days of agriculture. Herbicides have raised expectations of perfect weed control even higher, to the extent that tenant farmers who try to reduce their use of herbicides and suppress rather than eradicate their weeds may have problems with the landowner. Weed scientists and farmers in Iowa have identified this as one of the leading constraints to a more integrated approach to weed management in the state. That this is no trivial matter is seen in the 1992 census, which shows that 62 per cent of farmland in Illinois was rented.

Social pressure also has been reported to cause unnecessary pesticide use in Asian rice production:

> *A farmer who uses pesticides, even if the use is not warranted, is often perceived by his peers as up-to-date. In surveys and focus group interviews conducted in the Philippines and Thailand, many farmers admitted that seeing their neighbours spraying their fields often prompts them to spray as well, even though it is not necessary. In China and Vietnam, farmers who do not spray are often classified as uncooperative and may even be subjected to certain sanctions by village authorities.*
>
> (Escalada and Heong 1993)

Hangovers from the green revolution

The use of pesticides in food crops in developing countries was coupled with the spread of the green revolution. National crop intensification programmes promoted credit packages of the new high-yielding varieties and pesticides. The linking of the

Table 2.3 *Farmers' perceptions and actual values of rice yield reductions if no insecticides are applied – three provinces of the Philippines, 1980–91*

Iloilo		*Nueva Ecija*		*Camarines Sur*	
Farmer	*Actual**	*Farmer*	*Actual**	*Farmer*	*Actual**
32.7%	3.4%	35.1%	19.9%	44.3%	18.1%

Source: Compiled from Waibel (1986)
*Actual yield losses are mean differences between complete protection and untreated plot over 3 to 4 seasons, including non-significant differences

new varieties with pesticides was made stronger by severe problems of pest outbreaks – such as the brown planthopper in rice – with the first varieties, such as IR8. These varieties have been replaced with superior cultivars with resistance to many pests and diseases, but perceptions have not caught up. Rice farmers still tend to overestimate both the crop losses due to insects and the technical effectiveness of crop protection measures (see Table 2.3). Pesticides are consequently applied as insurance grounded in fear of loss.

Pesticide- and fertilizer-induced pests

The agricultural ecosystem is a complex of plant feeders, predators, parasites and other organisms that interact together to form rather stable units. Perhaps one or two key primary pests cause economic loss, while natural controls or agricultural practices prevent many other *potential* pests from reaching pest status. This applies to insects, diseases and weeds. When broad-spectrum insecticides are targeted at primary pests, there may be a resurgence of the primary pest after initially effective control because the natural enemies have also been suppressed by the chemical. Furthermore, other insects or mites may be unleashed from natural control and become more serious *secondary* pests than the original primary target because they have acquired resistance to the pesticides. An inventory of the 25 most serious pests in California in 1970 found that 17 of those pests were resistant to one or more classes of insecticides, and that 24 were either secondary pests or pest resurgences aggravated by the use of insecticides (Luck et al, 1977). Some of the primary targets of pesticide companies' insecticide screening, such as bollworms and spider mites, are in fact secondary pests.

In the decades that followed the introduction of synthetic pesticides, dramatic examples of a pesticide treadmill began to appear in which increasing frequency and dosage of pesticides were required every season to control resistant and secondary pests induced by the previous season's pesticide regime. Cotton provides numerous examples. In the US and Central America, spraying to control the primary pests – the boll weevil and the pink bollworm – led to the ascendance of secondary bollworm, whitefly and other pests, which developed resistance to most available pesticides even after 15–18 applications (sometimes 48–50 applications were recorded in Nicaragua). These pesticide treadmills in cotton led to the abandonment of cotton production in southern Texas and north-eastern Mexico and dramatic reductions of acreage in much of Central America and much more recently in California's Imperial Valley. Cotton production in north-eastern China is locked into a treadmill and looks very fragile. So-called resistance management in Chinese cotton has largely been a journey up a toxicological blind alley, exploiting the remaining sensitivity of chemical pesticides with aggressive mixtures of highly toxic products. Large-scale cotton failures in India have been due to resistant bollworms, which have spread from Andhra Pradesh to north and south India. Other famous human-induced pests are the brown planthopper in tropical rice, red spider mites and the poinsettia whitefly, each signalling the end of the line for chemically dependent strategies and the need for a rethink of pest management options.

Similarly, the use of herbicides causes shifts in weed populations. As one weed species is killed off, another proliferates in its place, often posing an equal or greater problem (Freemark and Boutin, 1995). In the 1980s there was an explosion of instances of resistance to the new generation of selective herbicides, and globally more than 180 species of weeds are now documented as resistance problems. In Australia, more than 3000 large wheat farms covering nearly one million hectares have weed 'biotypes' that are resistant to virtually all herbicides that can be used on wheat crops (Gianessi and Puffer, 1993). It has also been speculated that one reason for the massive increase in herbicide use between 1950 and 1980 could be that farmers and agronomists have been trying to compensate for the deterioration of the soil, which encourages weed infestation (Ghersa et al, 1994). These knock-on effects of pesticide and fertilizer use are very common. Excessive use of inorganic fertilizers is also known to leave crops more susceptible to some disease and insect pests, and the use of soil sterilants leaves crops more dependent on inorganic fertilizers.

Privatization of research or information

As public investments in agricultural research and advisory services decline around the world, farmers have turned increasingly to information from the private sector – especially agrochemical merchants – and to research conducted or sponsored by private companies, for their crop protection needs. The situation is worse, though by no means confined to developing countries, where extension workers often have a reputation of being underpaid, low-status generalists. In Indian Punjab, 60 per cent of farmers look to pesticide dealers for information, compared with 6 per cent who look to agricultural universities or extension services (Mehrotra in Ciba Foundation, 1993, p95). In China most extension workers have to obtain at least 30 per cent of their salary and most of their operational budget by free marketing – usually selling pesticides. The situation is made worse by advertisements which play on farmers' fears of crop loss, and by the fact that many pesticide salesmen are paid commission based on sales volume. Integrated approaches to crop production and protection cannot be built on a relationship in which all problems have product solutions, just as a systemic approach to health care cannot be led by pharmacists.

In the first section of this chapter, we have presented an historical account of the factors which have caused farming's increased reliance on chemical pesticides over the last 50 years. We conclude that pesticides have had more to do with farmer survival than with feeding the world. They are a farmer's tool in securing a market for his food and fibre produce, in an extremely competitive global market, by meeting the demands of price and quality imposed by the gatekeepers of the agrifood system, the processors, the packers and the retailers. Past developments are poor indicators of future trends, and there is strong evidence that many of the historical driving forces are changing (see Chapter 5). Before we look at alternatives to pesticide dependence, we will briefly examine some of the consequences of farming's reliance on chemical pest control.

The Consequences of Pesticide Dependence

The consequences of pesticide use – on the health of farming, the health of people, the health of the environment, and on the health of our political establishments – is a hornet's nest of controversy. 'Scientific' studies can be selected to support a very wide range of interpretations, and it is difficult to comment on these issues without accusations of scare-mongering, muck raking, gross complacency, or pandering to vested interests. Much has been covered elsewhere in the literature (Conway and Pretty, 1991; Dinham, 1993; Pingali and Roger, 1995) and will be analysed in economic terms in Chapters 4 through 5. Our intention is to summarize recent information and to focus on two aspects of pesticide dependence that receive scant attention – the consequences of pesticide production, transport and storage, and the political consequences of the pesticide industry's entrenched position.

Like the debate over the rationale and justification for pesticide use, agribusiness and its allies in the scientific establishment take a far rosier view of the consequences and side-effects of pesticides. Reassurances usually centre upon food and water and the negligible risks from pesticide residues:

> *A 150-pound adult would have to eat 3000 heads of lettuce each day for the rest of his or her life to ingest the amount of pesticide that is found to cause health problems in laboratory mice.*

> *The US Environmental Protection Agency well water survey concluded that 96% of America's 10.5 million rural and community drinking water wells are completely free of detectable traces of any of the 126 pesticides for which the wells were tested.*

In keeping with these messages from the US Agricultural Retailers Association (1996), the anthropocentric aspects of pesticide dependence – residues in food and drinking water – consume the lion's share of the pesticide debate and research investments. We briefly put this debate into context before moving on to the ecological consequences of pesticide use for farmworker and ecosystem health.

Food and drinking water

Fruits and vegetables grown in the US and EU – while likely to contain some detectable residues (EWG, 1995) – usually have less than 1 per cent of samples in excess of the maximum residue limits (MRLs). The consensus among most scientists is that, with some notable exceptions, pesticide residues in food pose very little risk to consumers. High residues of pesticides on fruits and vegetables are a more serious problem in developing countries, where 15 per cent or more of samples of fruits and vegetables have residues in excess of MRLs set by the United Nations Food and Agricultural Organization (FAO). Citizens of many tropical countries – where DDT is still widely used for malaria control – are also exposed to high concentrations of organochlorine pesticides via fish (Kannan et al, 1995) and meat consumption. Large quantities of DDT and other organochlorines are still used in developing countries. High levels of DDT and lindane are still found in breast milk in India – a 1991 study found that infants in Delhi and the Punjab received roughly 12 times the allowable daily intake of DDT (Nair, cited in Repetto and Baliga, 1996). The concentration of organochlorines – DDE, PCBs and dioxins – is also alarmingly high in the breast milk of Inuits in the Canadian Arctic due to their dietary intake of fish, whale, seal, walrus and bear meat. The level of organochlorine exposure was correlated with greater incidence of inner ear infections, suggesting a link with immune dysfunction (Dewailly, cited in Repetto and Baliga, 1996).

The significance of pesticides in drinking water depends very strongly on how pesticide detections are interpreted in policy terms. In the EU, pesticides are dealt with as unwelcome contaminants of the water supply, and maximum permissible residue limits approach zero-tolerance (0.1 micrograms per litre). It is the consumer rather than the polluter who pays for adherence to this precautionary standard, which in the UK alone has required £1 billion ($1.6 billion) in new investments by water companies, as well as annual financing and operating costs of over £100 million (about £7 ($12) for every kilogramme of pesticide used; Pretty, 1998; Tye, 1997). In the US and Canada, standards for drinking water are health-based, so that while detections are also common, the costs for capital equipment are considerably less. Communities which depend on surface water and reservoirs rather than on groundwater are more likely to be exposed to herbicide contaminants in their drinking water, especially in the summer months. The US–EPA estimates that between 50,000 and three million people may be drinking water contaminated with herbicides above EPA standards (*Pesticide and Toxic Chemical News*, 26 June 1996; see also Wiles et al, 1994).

It is worth noting that, in contrast to the portrayed brave new world of biorational chemicals and genetically engineered seeds, we are still living with a cocktail of organophosphates, organochlorines and other 'established' chemistry in an infinite range of combinations. The potential role of hormone-mimicking chemicals, including pesticides, in the dramatic decline in male reproductive health over the past 30 to 50 years (Toppari et al, 1996; Arnold et al, 1996), has added weight to a precautionary rather than health-based approach to pesticide regulation.

Farmers and farmworkers

Fatalities at work in Europe and North America due to pesticides are very rare – one a decade in the UK, and eight a decade in California – and there are many other more common causes of death on the farm. But increasing attention is being paid to the long-term *chronic* health impacts of pesticides. A study in the US state of Minnesota (Garry et al, 1996) showed that the offspring of pesticide appliers had significantly higher risks of birth abnormalities, specifically urogenital and musculoskeletal defects, as well as alterations in sex ratio. About 25 per cent of farm labour in the US is performed by children (Farmworker Justice Fund, 1990). A study in New York State (Pollack, 1990) found that over 40 per cent of interviewed children had worked in fields that were wet with pesticides, and a similar proportion had been sprayed while in the fields. The US–EPA have used data from California to estimate that nationwide there are between 10,000 and 20,000 physician-diagnosed pesticide illnesses and injuries among agricultural workers (US–EPA, 1992), and Blondell (unpublished) believes that these figures should be doubled to get a realistic picture due to the significant underreporting of farmworker poisonings.

In developing countries, mortality and illness due to pesticides are much more common than in OECD countries relative to the amount of pesticides used (see Box 2.1). The *British Journal of Industrial Medicine* reported that although 75 to 80 per cent of agricultural chemical use occurs in developed countries, developing countries account for more than 99 per cent of all deaths from pesticide poisoning. Lack of legislation, widespread misunderstanding of the hazards involved, poor labelling and the discomfort of wearing full protective clothing in hot climates greatly increase the hazard to agricultural workers. A study of farmers in the Philippines found that those exposed to

Box 2.1 Pesticide poisoning in selected developing countries

- In Malaysia and Sri Lanka, 7 to 15 per cent of all farmers experience poisoning at least once in their lives.
- In Thailand, a survey of 250 government hospitals and health centres revealed that some 5500 people were admitted for pesticide poisoning in 1985 alone, of whom 384 died.
- In the Philippines, 50 per cent of rice farmers have suffered from sickness due to pesticide use.
- In Latin America, 10 to 30 per cent of agricultural workers show inhibition of the blood enzyme, cholinesterase, which is a sign of poisoning by organophosphate pesticides;
- In Venezuela, 10,300 cases of poisoning with 576 deaths occurred between 1980–90;
- In Brazil, 28 per cent of farmers in Santa Catarina say they have been poisoned at least once; and in Paraná, some 7,800 people were poisoned between 1982–1992.
- In Egypt, more than 50 per cent of cotton workers in the 1990s suffer symptoms of chronic pesticides poisoning, including neurological and vision disorders.
- In China, 42,800 new cases of pesticide poisoning were reported in 1994, including 3900 fatalities – many were said to be victims of homemade cocktails marketed illegally, and some 30 per cent of products were unlicensed by authorities.

pesticides had statistically significant increased eye, skin, lung and neurological disorders (Rola and Pingali, 1993). Moreover, many pesticides known to be highly hazardous and either banned or severely restricted in industrialized countries are widely available. At least 150,000 tonnes of hazardous pesticides – including 11,000 tonnes of products banned in the US and 10,000 tonnes of severely restricted pesticides – were exported from the US between 1992 and 1994, mostly to developing countries (*Pesticide & Toxic Chemical News*, 29 May 1996, pp 16–17). The FAO estimates that developing countries are holding stocks of more than 100,000 tonnes of obsolete pesticides, 20,000 tonnes of which are in Africa. Most of these pesticides are left over from pesticide donations provided by foreign aid programmes. These obsolete stocks have been called 'potential time bombs' by the head of the FAO's Plant Protection Service (*FAO News and Highlights*, 1996). Lastly, we should reiterate that pesticides have affected the lives and livelihoods of farmers and farmworkers, both in terms of freedom from drudgery and as one of the suite of technologies which have 'released' millions of people from the land over the past 50 years.

Ecosystems

Agriculture is the dominant land use pattern across the world, and when terrestrial, aquatic and marine ecosystems show signs of stress – perhaps in the precipitous decline of bird or amphibian populations – the finger of blame inevitably turns to pesticides. After all, 2.6 million tonnes of pesticide active ingredients are applied around the world each year (EPA, 1997), some of which find their way to streams and oceans; 630–1000 tonnes of triazine herbicides are flushed annually from the Mississippi Basin to the Gulf of Mexico (*Pesticide and Toxic Chemical News*, 3 April 1996, p15). But to link cause and effect stretches epidemiology to its limits; the more we know about the impacts of pesticide mixtures and other toxins at minute concentrations, the more questions are raised. For instance, it may be found that bees are killed by residues of an insecticide, but their homing flight is disrupted by minute doses. An aquatic plant may not be directly affected by traces of a herbicide, but the subtle balance between the plant and the herbivores that live on it is tipped in favour of noxious species. Traces of pesticides in seawater may be found to be non-toxic to fish fry in test conditions, but in the ocean those traces may be concentrated to a lethal mixture within a surface microlayer. Declines in birds may be due to direct toxicity of pesticides, or through declines in insect or plant food and nesting cover, or due to other effects of agricultural intensification such as larger fields. We have to look for clues around the landscape, such as the remarkably higher nesting rates of birds and the abundance of non-pest butterflies on organic farms compared to conventional farms (Wilson, 1993; Chamberlain et al, 1996; Feber, 1996).

Silent Spring by Rachel Carson was primarily a testimony to the links between pesticide use and environmental health. It was a call for a precautionary approach to the release of persistent chemicals into the environment in the name of a war on pests. In the decades since its publication, we now know more about the atmospheric transport of pesticides and the importance of rainfall in the surface water accumulation of some herbicides; we know more about the hormone-mimicking effects of some pesticides in mammals and fish; we know more about the effects of pesticides on the immune systems of animals; and we know that the soil sterilant methyl bromide is responsible for 5 to 10 per cent of the thinning of the stratospheric ozone layer. But new knowledge of these effects has not given us any more confidence about the elegance or the precision of our armoury of chemical crop-protection weapons.

Impacts of production, transport, storage and disposal

With the horrific exception of the 1984 catastrophe at Bhopal, India, in which Union Carbide's pesticide-manufacturing plant leaked 42 tonnes of methyl isocyanate, killing thousands of residents, little attention has been paid to the sustainability of pesticide manufacture, formulation, transport and disposal. Almost all research and activism is focused on the risks of pesticide use.

The chemical industry has by far the largest level of releases from production sites of toxic chemicals to the environment. In the US, chemical companies reported nearly 400,000 tonnes of on-site releases, including nearly all of the nation's underground injection of waste (US–EPA 1996). A survey of the US Toxics Release Inventory for 1994 shows that agrochemical production facilities of the largest pesticide companies produced nearly 70,000 tonnes of production-related waste such as acids and solvents, though companies are reporting big improvements in production efficiency, with Monsanto reducing releases to the air by 90 per cent. An increasing amount of waste is being recycled or incinerated, but 4200 tonnes were released on-site to air, land or water, pumped underground or disposed of off-site.

Pesticide production, formulation or repackaging plants and pesticide dumps on the US–EPA's national priority list of contaminated 'Superfund' sites are mainly expensive legacies of the organochlorine era. Such pesticide-contaminated sites in 17 US states have estimated cleanup costs of $850 million. Some of these sites have soil, surface water and/or groundwater contaminated with lindane, toxaphene, DDT, chlordane, heptachlor, mirex, dieldrin and endrin, as well as dioxins from the production of

Table 2.4 *The risks of pesticide production, storage and transport – a selection of incidents, 1984–97*

Date and Place	*Company*	*Incident*	*Impacts*
December 1984 Bhopal, India	Union Carbide	42 tonnes of methyl isocyanate leaks from pesticide plant	Over 4000 people killed and up to 500,000 injured, 50,000 partly or totally disabled. Union Carbide settled in 1991 for $470 million: Indian government had asked for $3.3 billion. Worst industrial accident in history.
November 1986 Basle, Switzerland	Sandoz	Warehouse fire; 1300 tonnes of pesticides, dyes, solvents, etc released to atmosphere, soil, ground water and Rhine River	Rhine biota severely damaged for at least 400 km.
May 1991 Cordoba, Mexico	Veracruz National Agriculture Company (ANAVERSA)	Fire and explosion at pesticide warehouse and mixing plant in densely populated area – >31,500 litres of methyl parathion, paraquat, 2,4-D, and penta-chlorophenol burned or spilled from the site	759 diagnosed pesticide poisonings, 55% demonstrated persistent neurological damage. No deaths.
July 1991 California, US		Train derailment spills 75,000 litres of metam sodium into the Sacramento River	Virtually all aquatic life killed along 42 miles of upper Sacramento River. Over one million fish killed. 705 people report health effects.
February 1993 Escaguey, Venezuela		Truck carrying three tonnes of pesticides drives off a mountainous slope, spilling multiple chemicals, including parathion, paraquat, aldrin, and DDT	Over 1000 local residents poisoned, 78 severely. Poisonings from direct exposure and toxic cloud after containers were burned.
Winter 1993–4 NW Europe	Ciba-Geigy	Loss of container containing seed treatment pesticides from freighter *Sherbro* in English Channel	Thousands of sachets washed up on French and Dutch beaches.
April 1994 Dallas, Texas, US		Truck carrying ten tonnes of aldicarb crashes and catches fire	17 people taken to hospital for possible exposure.
April 1994 Karachi, Pakistan	Ciba-Geigy	Warehouse explosion and fire. 85 tonnes of formulated pesticides destroyed, including monocrotophos, mancozeb, metalaxyl, cypermethrin and profenfos.	At least ten firefighters and one plant worker hospitalized.
May 1997 West Helena, Arkansas, US	BPS Inc	Warehouse explosion and fire triggered by fire in 300-lb bag of azinphos methyl	Three firefighters killed and 16 injured. Local residents treated with atropine.
October 1997 Birmingham, Alabama, US		Warehouse fire. 47,000 gallons of chlorpyrifos insecticide washed into lakes and streams.	Massive fish and wildlife kills along 26 miles of Village Creek.

Source: Compiled from Pesticide Action Network PANNIS database

2,4,5–T, and require long-term pumping and treating of contaminated groundwater and soil excavation with thermal treatment. Site contamination may also be associated with the use of old products on crops. Over a period of 50 years, Washington state orchardists used up to 16 million kilogrammes a year of lead arsenate on between 28,000 and 49,000 hectares of apples. Clean up – at $50,000 to $75,000 per hectare – is required if the sites are developed for housing.

Increasingly, production of the older and often highly toxic products occurs in developing countries, close to the major markets. The large multinationals and smaller local companies are engaged in this southward shift in production capability (Dinham, 1996). From the multinationals, examples include production plants for monocrotophos and paraquat in China which are due to come on-stream in 1996 for Novartis and Zeneca, with capacities of 3000 to 6000 tonnes per year. India's pesticide production capacity stands at 86,000 tonnes of over 60 technical grade products, manufactured by more than 125 companies. Production in China is growing rapidly, standing at 250,000 tonnes of active ingredient in 1995, equivalent to 1.5 million tonnes of formulated product; 30 per cent of this production ends up as waste because of inferior quality (*Pesticides News* 34, p18).

Between production site and farmer, both environmental and health impacts are inevitable when accidents and fires occur during storage and transport. A selection of incidents in the decade since the Bhopal catastrophe are listed in Table 2.4. At the point of sale – the thousands of dealerships scattered through agricultural areas – water and soil may be severely degraded. In the state of Iowa alone, up to half of the 1500 dealerships are estimated to have groundwater contamination exceeding the state's cleanup guidelines. In that US state, the scope of the problem of assessing, monitoring and remediating at 28 agrochemical dealer sites is estimated to be between $50 and $100 million. These costs could force the closure of many of these businesses, with 8500 jobs at risk (Gannon, 1992).

Impact on government and governance

Conflict-ridden areas such as pesticides in agriculture attract heavy state intervention – labelled by critics on both left and right as an 'ecocracy' or 'toxic bureaucracy' fed by the 'environmentalist–regulator complex' engaged in self-perpetuating gridlock. The introduction of regulations and restrictions makes the system lock up and stop evolving, preventing technological and social advances (Akimoto, 1995). The governmental management and regulation of pesticide risks has proven to be an increasingly expensive exercise in terms of financial and human resources in both the public and private sector.

An analysis by Benbrook (1996) estimated that between 1971 and 1995, public sector expenditure on the regulation and management of pesticide risks in the US totalled almost $7 billion in 1995 dollars, equivalent to 7.4 per cent of sales value. This figure is predicted to be over $14 billion for the next 25 years (1995–2020). In addition, the pesticide industry, the food industry and farmers spent $10 billion on compliance, residue testing and so on from 1970–95. By comparison, the US Department of Agriculture's (USDA) budget for IPM research in 1995 was $192 million (Benbrook, 1996, p213). In short, pesticide regulation currently costs American industry and taxpayers more than a billion dollars a year, money which should be helping to manage pests and ecosystems rather than pesticides (Groth, 1996).

The huge investments of chemical companies into pesticide production also create a powerful vested interest to be protected through lobbying and sponsorship of 'friendly'

Table 2.5 *Contributions by pesticide-producer political action committees (PACs) to members of the US Congress, January 1989–April 1996*

Company or Organization	*PAC Name*	*Contribution to US Congress, 1/89–6/94*	*Contribution to US Congress, 11/92–4/96*
American Crop Protection Association	American Crop Protection Association PAC	$21,750	$14,150
American Cyanamid	American Cyanamid Good Government Fund	–	$38,471
	American Cyanamid Employee Fund	$24,500	$83,300
Bayer	Bayer Corporation PAC		$61,000
Ciba-Geigy	Ciba Employee Good Government Fund	$447,488	$193,363
Dow	Dow Chemical Company Agricultural Executive PAC	$110,150	$71,150
	Dow Chemical Company Employees' PAC	$180,250	$60,750
DuPont	Dupont Good Government Fund	$141,681	$122,714
FMC	FMC Corporation Good Government Program	$726,433	$332,700
Monsanto	Monsanto Citizenship Fund	$254,780	$127,305
Rhone-Poulenc	Rhone-Poulenc Inc PAC	$190,700	$79,500
Rohm and Haas	Rohm and Haas Employees Association for Better Government	$70,794	$42,376
Sandoz	Sandoz Agro Inc Patriot Committee	$25,950	$30,400
Zeneca	Zeneca Inc PAC	$49,100	$127,247

Source: Wirth and Schima (1994), Davies et al (1996)

science in support of pesticide-based agriculture and in opposition to bans, restrictions, taxes, use-reduction targets or incentives for low-input agriculture. They were evident in the pesticide industry's response to *Silent Spring* – in which Rachel Carson was branded as hysterical, in league with the communists and 'unscientific' – and are still in evidence today. These activities can continue to corrode the responsiveness of democratic government and science to wider societal interests. This is recognized in high places. US Vice President Al Gore (1994) wrote recently of the 'twin contaminations of special interest money and influence' and that 'hardliners within the pesticide industry have succeeded in delaying the implementation of protective measures called for in *Silent Spring*'. 'Cleaning up politics', he writes, 'is essential to cleaning up pollution'.

In the US, political action committees (PACs) are used by interest groups to funnel campaign contributions to candidates who might produce political favours. Since Congress members depend on PACs for 27–40 per cent of their campaign funds, they have become more beholden to PAC constituent groups (Gardner, 1995, p189). Data are available on PAC contributions from companies that produce pesticides to members of the US Congress (see Table 2.5). These contributions are not, of course, all targeted at pesticide legislation, especially for companies such as FMC for whom pesticides are a small proportion of their business. Nevertheless, the Environmental Working Group found that supporters of pesticide industry-friendly legislation were likely to receive five times as much from these PACs as supporters of a reform measure upheld by consumer and environmental groups (Wirth and Schima, 1994).

Another example is the dilution of the Clinton Administration's June 1993 commitment to 'reducing the use of pesticides and to promote sustainable agriculture'. This set a goal for a 75 per cent implementation of IPM by 2000 and promised specific pesticide use-reduction goals for various agricultural segments. The response by agribusiness was hostile, as Benbrook (1996, pp 209–10) recalls:

> *Chemical industry representatives opposed any mention of 'use reduction' as a goal, favouring instead risk reduction based on 'sound science'... Registrants also argued strongly that it was inappropriate and possibly illegal for EPA to embrace use or risk reduction goals since all registered pesticides have been determined by the agency to pose acceptable risks when used in accord with labels.*

The reduced-use approach was described as a non-starter by the American Crop Prtoection Association's director of scientific and regulatory affairs (*Agrow* 202, p14). So it turned out to be. Between August 1994 and April 1995 the EPA dropped first references to the 75 per cent IPM goal, and then references to the need to reduce use or risks.

The costs to industry of public relations campaigns are significant. The Agricultural Council of America launched in 1990 a $50 million Food Watch campaign to 'restore public confidence in the safety of the food supply'. A recent $1 million Shared Values initiative to extend Food Watch with children as the campaign's 'key visual' (*Feedstuffs*, 3 June 1996, p30) was abandoned in mid 1996 due to lack of industry support. The American Crop Protection Association supplies free educational material for schools, featuring Benny Broccoli and his Buddies, describing how fungicides help plucky plants to fend off fungi and other lessons. Meanwhile, the National Agri-Marketing Association – which includes American Cyanamid, Bayer, DuPont and Monsanto – is seeking funding for its United Voice campaign to 'bolster confidence in and acceptance of high-quality, abundant, affordable, safe US agricultural products'. Donations to conservative think tanks such as the Hudson Institute, Accuracy in Media (AIM) and Consumer Alert also run into hundreds of thousands of dollars – between 1992 and 1995, agrochemical manufacturers contributed around $150,000 to the Hudson Institute, publisher of *Saving the Planet with Pesticides and Plastic* (Public Citizen, 1996). Environmental organizations also use the same tools to get their voices heard. The Sierra Club spent almost $7.5 million trying to influence the outcome of US congressional races in 1996 – during an election campaign where total spending exceeded $1.8 billion.

The influence of industry also appears to have a marked impact on the outcome of scientific studies to test the safety of chemicals. In a recent analysis of 161 studies of the carcinogenic properties of four chemicals, including two herbicides, Fagin et al (1997) found that of 43 industry-funded studies, only six returned results unfavourable to the chemicals involved. But in the 118 studies conducted by non-industry researchers, 71 were unfavourable. We should make it clear that vested interests of chemical industries do not make their presence felt only in the OECD. Pesticide industries are protected in a number of developing countries, such as India and China, in ways which can impede the adoption of safer pesticide alternatives.

The response of government bureaucracy and industry to the environmental conflict over pesticides has a large opportunity cost. Money spent on monitoring pesticide residues in food and water could have been invested in research on low-chemical, biologically intensive forms of agriculture. Research is locked into wasteful cleaning up after industrial agriculture rather than exploring new agricultural innovations. The dialogue becomes 'scientized' into a debate on the pros and cons of pesticides rather than on the technical, political and social means to achieve agricultural sustainability. This will be explored further in Chapter 6.

To summarize this section, it is clear that the consequences of pesticide use extend way beyond the issues of pesticides on food or in water. To couch the argument in terms of how many thousand heads of lettuce we must eat or gallons of tap water we must drink to risk ill health from pesticides is to miss the major point of direct pest control. This technology does not appear to be sustainable in its own terms – of controlling pests – when used as a primary pillar of agricultural production. *The technology is not durable.* Pesticide usage is responsible for a decrease in its own efficiency. Reliance on pesticides always produces 'revenge' effects (Tenner, 1996) of resistance or secondary pests which are even more difficult to control than the primary target. The heavier that reliance (such as control of 'exotic' pests, or in most cotton-producing areas worldwide), the quicker revenge effects become apparent.

Many of the impacts of pesticides on human and environmental health which are being measured in the 1990s concern products such as the organochlorines DDT, DBCP, lindane and endrin which are now banned in many countries, and with older, highly toxic and non-selective organophosphates. The pesticide industry and its regulators can rightly claim that most of these products now have safer, low-rate alternatives, and that the health impacts and chronic ecosystem degradation experienced in the past 30 years are now shifting towards a more positive ratio of benefits to costs. There are two limits to this line of argument. Firstly, companies continue to defend and produce their older products, and the manufacture of these products is being shifted towards developing countries where the bulk of sales are anticipated in the years ahead (Dinham, 1996). Secondly, history is teaching us a cautionary lesson about the subtle effects of chemicals in the environment and in the body. The chronic effects of pesticides, such as DDT, on the endocrine system of humans and wildlife took 50 years to come to light, and it is naïve to expect that the batteries of tests now required to register a pesticide can anticipate all future chronic health and environmental impacts of the 'safer' products. This is not to tar all pesticides with the same brush, and it is not to equate the risks of pesticides with more serious threats such as overfishing and nutrient enrichment in marine systems and the carbonization of the atmosphere. But before we load the environment with more foreign chemicals or build huge plants to produce them, precaution should steer us away from substitution and towards reduced dependency on pesticides.

REDUCED DEPENDENCE ON PESTICIDES

We have seen in the first section of this chapter that pesticide use has been a factor allowing the revolutions of *specialization* and *intensification* in farming. To be able to compete in a global market on quantity, quality and especially price, pesticides seem to be part of a farmer's survival tool kit, despite their obvious limitations. In the next section we draw attention to some of the consequences of this reliance, which may be fundamentally unsustainable in terms of impacts on farmer welfare, ecosystem health and biodiversity. Our analysis shows the hollowness of the agribusiness claim to feed the world, but at first glance it also suggests that farmers are locked into the agrochemical–biotechnological treadmill by global trends in economics and commerce. This would indeed be the case if chemicals were essential for agricultural intensification. But in this final section, we look at the predictions made if pesticides were to become a less integral part of modern farming, and compare those predictions with evidence of viable alternatives to agricultural development for both industrial and subsistence farming, based on biological and management-intensive systems.

The language of loss

In the 35 years since the publication of *Silent Spring*, the response by industry and some politicians and scientists to agriculture's reliance on chemical pesticides has remained distressingly entrenched in what Bromley (1994) calls 'the language of loss'. The debate is framed in terms of the costs of pesticide bans: the highly unlikely scenario of deleting pesticides from the existing mix of agricultural practices. This tactic was used in the DDT debate by Earl Butz, by the Nobel prize-winning 'father of the green revolution' Norman Bourlaug, by American Cyanamid's Robert White-Stevens (Hynes, 1989, p18), and by Monsanto, who rushed out a hostile parody of *Silent Spring* called *The Desolate Year* to coincide with the launch of Carson's book, portraying an impoverished and infested US in an imaginary year when no pesticides were used in agriculture or households. And yet Rachel Carson herself never called for a ban on pesticides. She had instead advocated the 'other road' of biologically based pest-control methods 'based on understanding of the living organisms they [scientists] seek to control' and a minimum use of chemical pesticides. She was attacked for things she never said:

> *Anyone who has really read the book [*Silent Spring*] knows that I criticize the modern chemical method not because it controls harmful insects but because it controls them badly and inefficiently and creates many dangerous side effects in doing so.*
>
> (Rachel Carson quoted in Hynes, 1989, p42)

The tradition of attacking the straw man of total pesticide bans was continued by a research group at Texas A&M University led by Ronald Knutson, in work sponsored by a consortium which included DowElanco and Monsanto. And so it continues today. To see how the language of loss remains ingrained in the pesticide debate, we have to look no further, once again, than the US Agricultural Retailers Association's (1996) *Food for Thought* calendar:

> *Without pesticides and fertilizers, US farm exports would fall to zero. Our balance of trade would drop by more than $4 billion, and millions of people around the world would starve.*
>
> *An estimated one third of the world's food supply would be lost each year to weeds, insect pests and diseases if crop protection chemicals were not used. This is enough to feed about two billion people.*
>
> *Without pesticides and fertilizers, US consumers would spend 30 to 40 per cent of their income on food instead of the current 8 to 10 per cent, which is the lowest expenditure in the world.*

Alternative realities

What other 'realities' exist which challenge the language of loss? Does reduced-input farming really have to be reduced-yield farming, with inevitably higher food prices and expansion of cropland into wildlife habitat?

Yield reductions?

If pesticides are taken away from conventional farming systems without other changes in operations, there will obviously be severe yield losses due to competition from weeds and infestations of insects and fungi. But it seems foolish not to expect technological innovation by both public and private researchers to be restricted by certain chemical bans. And yet such 'treated versus untreated' comparisons are the norm when potential yield loss due to pests is calculated, and thus the amount of yield which is being saved by pesticides (Cramer, 1967; Knutson et al, 1990; Oerke et al, 1994). The Knutson study predicted crop losses of as much as 35–50 per cent if pesticides were not used in the seven crops studied, as well as food price inflation of over 10 per cent annually. Apart from ignoring technological responses, these data are further skewed by the fact that many conclusions are derived from pesticide field trials, in which conditions are manipulated to get a high pest infestation, for instance by planting susceptible crop varieties at a time of greatest pest pressure. Despite these serious shortcomings, the huge estimates of crop yields saved by pesticides go largely unchallenged (see Bender, 1994, pp 94–100).

There is good evidence that currently available best management technologies within conventional farming systems – minimum doses and precise application in response to economically justified infestations of pests – can lead to reductions in pesticide use in the order of 20 per cent. Such pest*icide* management is often misleadingly labelled integrated pest management (IPM). But pest management in the original sense of the term causes farming systems to intensify the action of those naturally occurring regulatory processes, placing the ecosystem rather than the pest and pesticide at the centre of the management approach (NRC, 1996, Rosset and Altieri, 1995). As Rosset and Altieri explain: 'The challenge is to identify the correct assemblage of species that will provide, through their biological synergisms, key ecological services such as nutrient cycling, biological pest control, and water and soil conservation.' Innovative farmers skilled in such ecologically based pest management give us a clearer idea of how much yield needs to be sacrificed when pesticide inputs are greatly reduced.

A classic example is tropical rice. Early attempts to introduce IPM in the Philippines through farmer training on crop scouting and treatment thresholds proved disappointing. Pesticide use actually *increased* as training raised farmers' awareness of pests. But when a new approach was tried in Indonesia which took an ecosystem approach with season-long 'field schools' that stressed crop health (including the ability of the rice plant to compensate for pest damage), dramatic reductions in pesticide input were achieved without yield loss. This approach has now been tried in many Asian countries with excellent results (see Table 2.6).

The rice example may give the impression that biointensive IPM does not need a large research investment. This is far from reality, especially for insect and disease control. The rice example was built on years of experiment station research and farmer field research. And for crops with strict cosmetic standards, high-yielding biointensive systems require a massive amount of research, especially in pest population ecology and soil quality. For example, the Biologically Integrated Orchard Systems (BIOS) project in California is showing how much research on cover crops is needed to manipulate pest, natural enemy populations and soil fertility to the benefit of the crop.

Very few coordinated programmes of biologically intensive farming have been conducted in OECD countries, where university research and government extension have concentrated on low-level 'efficiency'-type IPM. There are no figures available to compare with the FAO intercountry rice IPM programme. We have to seek out innov-

Table 2.6 *Impact of IPM training on farmers' pesticide application frequency and rice yields in Asia*

Country	*IPM-trained farmer pesticide application compared with untrained farmers*	*IPM-trained farmer crop yields compared with untrained farmers*
Bangladesh	n/a	+15%
China	–79%	+11%
India	–33%	+9%
Indonesia	–36%	+2%
Philippines	–50%	+2%
Sri Lanka	–26%	+23%
Vietnam	–57%	+8%

Source: FAO, Rome

ative farmers who are ahead of the curve, such as Ron Rossman of Iowa who is maintaining yields and improving soil quality and weed control as he moves away from pesticides and fertilizers. Rossman has saved $63,000 on herbicides since his transition to pesticide-free farming, which started in 1983 and was complete by 1997.

There are farmers like Rossman throughout the farming world. Their very survival outside conventional agriculture demonstrates their capacity to generate workable alternatives (Kloppenburg, 1991); improving competitiveness with conventional farming means that they can no longer be dismissed as lifestyle farmers or accused of exhausting the soil or fostering pests. These farmers are usually highly skilled managers. In row crop agriculture, they are distinguishable from their conventional neighbours by their longer rotations (with a legume such as lucerne or alfalfa) and incorporation of livestock. Their operations are also more likely to score higher on other indices of sustainable development, such as employment and contribution to rural economic development (Levins, 1996). Few farmers are willing or skilled enough to make a transition to high-yield management and biologically intensive agriculture. A recent Consumers' Union analysis puts only 4 to 8 per cent of the US farm acreage in their biointensive IPM category (Benbrook, 1996, p203). It is unlikely that polluter pays ecotaxes would do much to achieve the Consumers' Union goal of 100 per cent biointensive IPM by the year 2020. But if pesticides became much more severely restricted than now, farmer innovators such as Rossman give us an excellent example of the likely technological response of farming and agricultural research (public and private), and the correspondingly small yield penalty which would occur once a transition to these farming systems was complete.

The case for yield reductions when individual pesticides are cancelled can be even weaker. An interesting *ex post* case of predicted versus actual impacts of a pesticide cancellation is reported by Moore and Villarejo (1996) for the case of ethyl parathion in Californian lettuce. They show that the anticipated 25 per cent yield loss on which economists based predictions of a $14.3 million loss in consumer welfare (due to higher prices) never happened – there was no significant difference in yields with and without the chemical. After the infamous cancellation of Alar (daminozide) in 1989, the Washington State apple industry confounded the Cassandras by rebounding very rapidly, topping $1 billion sales for the first time in 1990.

Higher food prices?

On average, Americans spent about 11.4 per cent of total personal consumption on food in 1994, including around 8 per cent on food consumed at home, the lowest proportion in the world (Putnam and Allshouse, 1996). This compares to 18 per cent spent on food at home in Italy and 51 per cent in India. These statistics are a frequent justification for the role of chemicals in improving agricultural productivity.

The threat to this cheapness and abundance caused by restrictions on pesticides is likewise one of the most common laments of pesticide producers and retailers. Cheap food certainly does free up spending for other sectors of the economy, and as the Indian statistic shows, the price of food is of much greater significance to the welfare of the poor than of the rich. Japan and Switzerland – countries with higher per capita gross national product (GNP) than the US – spend over 20 per cent of consumption expenditures on food. At the high end of the income ladder, there is no correlation between the price of food and consumer welfare. But the US also has the highest inequality of wealth distribution in the OECD. The statistics on percentage of income spent on food are drastically distorted by high incomes of a very small number of families (Wessell, 1983, p124). They hide the fact that there is poverty among America's wealth – 33.5 million Americans who are officially poor, who spend 50 per cent or more of their after-tax income on food (Senauer et al, 1991, quoted in Clancy, 1993), even when over half of the US Department of Agriculture's budget between 1983 and 1991 was spent on food stamps for the poor. By March 1993, the number of Americans participating in the federal food stamp programme had risen to more than 27 million, about 10 per cent of the population (Kneen, 1993, pp 176–7). The United States has a child poverty rate far worse than any of the other 12 Western industrialized nations. While massive inequalities in wealth exist, the need for cheap food remains a powerful, though rather an uncomfortable justification for pesticide use.

The disingenuousness of the claim that the pesticide-dependent status quo provides cheap food is laid bare when we realize that advertizing and packaging soak up more of our food dollar than payments to farmers for the raw material (Ikerd, 1996). Lastly, we should note that cheap food may not be the bargain it appears. As discussed in Chapter 3, the full costs of producing it are not incorporated into the prices paid by consumers. Someone else pays the full costs of damage to the environment and to the health of farmers, farmworkers and consumers. If these external costs are internalized, then this changes the relative merits of different types of agricultural systems. Although a growing body of research shows that low-input sustainable systems can compete with modern agriculture, these modern systems prevail because of the greater subsidies they attract, and because the environmental and health costs are not internalized. The economic, environmental and political consequences of these subsidies are driving reforms of the CAP as well as pesticide regulations and taxation.

Food insecurity?

Global demand for food is projected to rise by up to 50 per cent in the next 15 to 25 years as a result of population growth and rising incomes. Most of the population growth will be in developing countries, many of them food-insecure. There is not much more land to be brought into cultivation, and the land available per capita has already fallen from about half a hectare in 1950 to just over a quarter of a hectare in 1990.

It has long been argued that pesticide use in food-exporting nations such as the US is essential to ensure that enough food is available for developing countries in the years ahead. Pesticide apologists used this argument when DDT was under scrutiny and continue the same arguments today. To concur with this rationale is to concur with the principle of a global free market as the best route to food security for less industrialized countries – countries buying the food they need at the best price on the international market. This new orthodoxy has replaced the orthodoxy of national self-sufficiency of the 1960s and 1970s. But trade-focused agricultural development, far from feeding the world, can severely disrupt local agriculture and trigger more unemployment and urbanization when local produce has to compete with the highly efficient production from Northern countries. Free trade in agricultural produce is tilted in favour of the US and EU where farmers are still highly subsidized, though subsidies are now more directed to income support rather than price support. But even without subsidies, agricultural imports from developed countries can undercut local produce, due to differences in factors such as labour-intensiveness, access to machinery, technology and advisory services, and transport infrastructure.

Food insecurity is largely a food-purchasing power problem and a political problem. People go hungry because they lack the means to buy or grow food. The current food surpluses of the Asia region simply mask the unmet needs of its vulnerable groups, which lack the land or money to ensure food security (Norse, 1994). As Lawrence Busch (1994) writes: 'Only by increasing household incomes of the poor in developed and developing nations will access to food be improved. Merely increasing the volume of food produced will not resolve the problems of the hungry.' For the least developed countries, food imports currently absorb more than one third of export earnings. The FAO estimates that the food import bills for 'low-income, food-deficit' countries will be 55 per cent higher in 2000 than in 1987–89. Dependence on imports means vulnerability to extreme price fluctuations, marginalization of small farmers, and ecological degradation.

> *Thus when countries with untapped agricultural resources provide food by importing more, they are effectively importing unemployment. By the same token, countries that are subsidising food exports are increasing unemployment in food-importing countries. This marginalises people, and marginalised people are forced to destroy the resource base to survive. Shifting production to food-deficit countries and to resource-poor farmers within those countries is a way of securing sustainable livelihoods.*
>
> Brundtland Commission (WCED, 1987, p129)

If a fraction of the subsidies which OECD countries spend on agricultural subsidies were ploughed into boosted aid for agriculture in developing countries, then the greater purchasing power of the poorest populations would go mainly on food. Examples of people-centred programmes of agricultural intensification covering nearly two million hectares across 22 developing countries, in which yields have improved without having to move to pesticide-dependence, are published in Pretty et al (1996).

Expansion into fragile land?

A fourth element of the language of loss is the assertion that pesticides are good for wildlife, because the yields they save in intensive agriculture allow wilderness and fragile ecosystems to remain unfarmed for the benefit of wildlife. The US Agricultural Retailers Association (1996) provides the following information bites:

> *One of the biggest benefits of modern farming is our ability to produce more food and fibre on our nation's best cropland. Fragile land can be taken out of agricultural production as we farm more suitable fields more intensively.*

> *The world cannot save its wildlife without high-yield agriculture and the careful use of farm chemicals because these practices allow crops to be produced on fewer and fewer acres, returning increasing amounts of land to wildlife habitat.*

The assertion that high-chemical input agriculture is good for wildlife assumes that:

- low levels of chemical inputs equate with low yields;
- wildlife and agriculture are incompatible;
- land-hunger and expansion into fragile zones are caused by low yields; and
- once a nation's food needs are met, fragile land is taken out of production.

We have seen that the first point will remain true only until legislative or market signals bring biologically intensive farming up to the level of today's innovators. For the second point, the mindset that food production is the only function of farmland, and that wildlife is compartmentalized into wilderness, ignores the evidence that farmland can support a far greater abundance and diversity of wildlife – with added benefits of landscape aesthetics, water quality and flood control – when farmers receive incentives and other signals from society which acknowledge agriculture's role in nature production as well as in food production. In the case of point three, history shows how low-input extractive farming systems have indeed driven the frontiers of agriculture across continents (Ponting, 1991) and that this process continues, especially in Africa. But modern-day land hunger is more a consequence of enclosure for export-oriented cash cropping rather than soil exhaustion. As for the last point, it is clear that once food demands are met by chemically intensive agriculture, it is far more likely that excess land will be used to produce exports or feedstocks for animal production and industry, rather than revert to prairie, forest or wetlands.

CONCLUSIONS

During the past 50 years, agricultural and rural development policies have successfully emphasized external inputs as the means to increase food production. This has produced a remarkable growth in the global consumption of pesticides, inorganic fertilizers, animal feedstuffs, and tractors and other machinery. These external inputs have, however, substituted for natural control processes and resources, rendering them more vulnerable. Pesticides have replaced biological, cultural and mechanical methods for controlling pests, weeds and diseases; inorganic fertilizers have been substituted for livestock manures, composts and nitrogen-fixing crops; information for management decisions comes from input suppliers, researchers and extensionists rather than from local sources; and fossil fuels have substituted for locally generated energy sources. What were once-valued local resources have often now become waste products.

The basic challenge for sustainable agriculture is to make better use of available physical and human resources. This can be done by minimizing the use of external inputs, by regenerating internal resources more effectively, or by combinations of both. This ensures the efficient use of what is available and ensures that any dependencies on external systems are kept to a reasonable minimum. A more sustainable agriculture, therefore, is any food production system that systematically pursues the following goals:

- a thorough incorporation of natural processes such as nutrient cycling, nitrogen fixation and pest–predator relationships into agricultural production processes, ensuring profitable and efficient food production;

- a minimization of the use of those external and non-renewable inputs with the potential to damage the environment or harm the health of farmers and consumers, and a targeted use of the remaining inputs used with a view to minimizing costs;
- the full participation of farmers and other rural people in all processes of problem analysis, and technology development, adaptation and extension;
- a more equitable access to productive resources and opportunities, and progress towards more socially just forms of agriculture;
- a greater productive use of local knowledge and practices, including innovative approaches not yet fully understood by scientists or widely adopted by farmers;
- an increase in self-reliance amongst farmers and rural people;
- an improvement in the match between cropping patterns and the productive potential and environmental constraints of climate and landscape to ensure long-term sustainability of current production levels.

Sustainable agriculture seeks the integrated use of a wide range of pest, nutrient, soil and water management technologies. It aims for an increased diversity of enterprises within farms combined with increased linkages and flows between them. By-products or wastes from one component or enterprise become inputs to another. As natural processes increasingly substitute for external inputs, so the impact on the environment is reduced.

There are several misconceptions about sustainable and regenerative agriculture (Pretty, 1995). The most common characterization is that sustainable agriculture represents a return to some form of low-technology, 'backward' or 'traditional' agricultural practices. This is manifestly untrue. Sustainable agriculture implies an incorporation of recent innovations that may originate with scientists, farmers or both. Another misconception is that sustainable agriculture is incompatible with existing farming. For the development of a sustainable agriculture, there is a need to move beyond the simplified thinking that pits industrialized agriculture against the organic movement, or the organic movement against all farmers who use external inputs. Sustainable agriculture represents economically and environmentally viable options for all types of farmers, regardless of their farm location and their skills, knowledge and personal motivation. It is also commonly stated that low-external input farming can only produce low levels of output. This is untrue. Many sustainable agriculture farmers show that their crop yields can be better than or equal to those of their more conventional neighbours. This offers the opportunities for growth for communities that do not have access to, or cannot afford, external resources. Either way, this means that sustainable farming can be compatible with small or large farms, and with many different types of technology.

We have seen in the first section that the bulk of pesticide use has surprisingly little to do with 'feeding the world' or keeping famine at bay, or even with killing pests. The largest markets for pesticide manufacturers – weed control with herbicides in OECD grain crops – are means to substitute capital for labour, equipment services and management in the low-margin, highly leveraged farming operations which have become the norm of OECD agriculture. Furthermore, much of this pesticide use is for growing *animal feed* for intensive livestock production in the North rather than for direct human nourishment. Even the big markets of developing countries are primarily for cash grain exports (such as the huge soybean, maize and wheat farms of Brazil and Argentina) for the international cash grain-livestock complex, or for cotton production. Pesticides are used as a crutch to prop up other inherently unsustainable farming practices.

Pesticides are also an integral part of a global food system in which production is increasingly *distanced* from consumption, necessitating high levels of inputs to grow crops out of season and to preserve quality in transit and storage. But a tendency to represent pesticide use as a conspiracy by rapacious corporations to lock farmers into a treadmill of pesticide dependency is also fallacious. Virtually all studies show that the marginal benefits of pesticides exceed marginal factor costs by several times or more (Carrasco-Tauber, 1990). This contradicts much anecdotal evidence that pesticides are overused in agriculture. On the contrary, farmers appear to *underuse* pesticides in simple cost-benefit terms; cash grain farmers in the US midwest, for example, are very reluctant to apply insecticides against corn borers, even though the 'economic threshold' of damage may have been exceeded. Instead, it was other powerful and well-established trends to which farmers were responding when they switched from mechanical to chemical weed control. Pesticide use – especially in Europe and Japan – has also been stimulated by artificial price signals from high-production subsidies. But although pesticides work to the initial benefit of individual farmers – allowing simplification of management over more acres – they work against the interests of *farming*. Pesticides are among a raft of technologies which have contributed to an inexorable shift in agricultural activities away from the farm. The share of farming in the value added to agriculture has been in steady decline since at least 1920.

Of course, some pesticides are used to protect yields of food crops in food-deficient countries, such as rice and vegetable production in south and south-east Asia. But this is a trivial market by the standards of the global pesticide trade, and cannot justifiably be used as a corporate raison d'être. Furthermore, a large slice of the tropical rice insecticide market is rightly viewed as a hangover from green revolution input packages, and that a well-managed, biologically intensive rice ecosystem needs very little insecticide inputs. Even the arguments about productivity and lower food prices appear to be spurious if we reevaluate this justification in the context of income inequities, and compare national statistics of food expenditures, quality of life and malnutrition.

In the second section we observed that scientific concern about subtle, chronic and often indirect effects of pesticides – which *Silent Spring* brought to the fore – are once again dominating the debate about human health impacts. It is unlikely that the concerns about reproductive and immunological effects will go away. And it is clear that inhabitants of developing countries have a disproportionately high exposure to pesticide risks, either during farm work or consumption of tainted fruits and vegetables. In the third section we saw that living with far less pesticides is a potential reality which has obvious consequences for our tolerance of the risks associated with pesticide use. But we are hindered by the language of loss employed by agribusiness and a paradigm of managing products rather than populations, which puts pesticides rather than ecosystems at the centre of the pest management approach. What is clear is that while pesticides have created opportunities – for example, for out of season fruits and vegetables in the supermarkets of the OECD – they have also precluded other opportunities. Huge resources have been invested in their development, in their monitoring and in their regulation – investments that could have yielded technologies that are more acceptable to the public and have a less arrogant and more precautionary approach to the 'chemicalization' of the environment. Now that these investments have been made, they obviously represent a major vested interest for producer, user, public scientist and regulator.

So, pesticide use appears to be a *symptom* of restructuring in the agrifood system and in the wider global economy, as well as a contributor to agriculture's unsustainability. The blame for many of the underlying forces cannot be laid on the producers of pesticides or their users. But neither can the production and use of pesticides be redeemed by attaching the label of sustainable development to this business.

REFERENCES

Akimoto Y (1995) 'A new perspective on the eco-industry'. *Technology Forecasting and Social Change* 49, pp 165–173.

Allen P and Sachs C (1993) 'Sustainable agriculture in the United States: engagements, silences and possibilities for transformation' in P Allen (ed) *Food for the Future: Conditions and Contradictions of Sustainability*, pp 139–167. John Wiley and Sons, New York.

Anderson K and Blackhurst R (eds) (1992) *The Greening of World Trade Issues.* Harvester Wheatsheaf, Hemel Hempstead, UK.

Ange S (1996) 'Economic analysis of crop protection policy in Costa Rica'. Pesticide Policy Project Publication No 4, Institute of Horticultural Economics, Hannover, Germany.

Babcock B, Lichtenberg W and Zilberman D (1988) 'The impacts of damage control in the quantity and quality of output: pest control in North Carolina apple orchards'. Working Paper no 88–39, University of Maryland, Maryland, US.

Balthasar CW (1995) *Free Trade, Farmers, and the Killer Tomato.* Santa Monica, CA, US.

Bender J (1994) *Future Harvest: Pesticide-Free Farming.* University of Nebraska Press, Lincoln, NE, US.

Benbrook C (1996) *Pest Management at the Crossroads.* Consumers Union, Yonkers, NY.

Blondell J (1997) 'Epidemiology of pesticide poisonings in the US with special reference to occupational cases'. *Occupational Medicine: State of the Art Reviews* 12(2) pp 209–220.

Bromley DW (1994) 'The language of loss: or how to paralyze policy to protect the status quo'. *Choices*, third quarter 1994, p31.

Busch L (1994) 'The state of agricultural science and the agricultural science of the state'. In A Bonanno et al (eds) *From Columbus to ConAgra: the globalization of agriculture and food*, pp 69–84. University Press of Kansas, Lawrence, Kansas.

Buttel FH (1993) 'Socioeconomic impacts and social implications of reducing pesticide and agricultural chemical use in the United States'. In D Pimentel and H Lehman (eds) *The Pesticide Question: Environment, Economics, and Ethics,* pp153–181. Chapman and Hall, New York and London.

Buttel FH (1994) 'Agricultural change, rural society and the state in the late twentieth century'. In D Symes and AJ Jansen (eds) *Agricultural Restructuring and Rural Change in Europe*, pp 13–31. Agricultural University, Wageningen, The Netherlands.

Carson R (1964) *Silent Spring.* Houghton Mifflin, New York.

Chamberlain D, Fuller R and Brooks D (1996) 'The effects of organic farming on birds'. *Elm Farm Research Centre Bulletin* 21, pp 5–9, Jan 1996. Elm Farm Research Centre, Newbury, UK.

Ciba Foundation (1993) *Crop Protection and Sustainable Agriculture.* John Wiley & Sons, Chichester, UK.

Clancy KL (1993) 'Sustainable agriculture and domestic hunger: rethinking a link between production and consumption'. In P Allen (ed) *Food for the Future: Conditions and Contradictions of Sustainability.* John Wiley & Sons, New York.

Clark J and Lowe P (1992) 'Cleaning up agriculture: environment, technology and social science'. *Sociologia Ruralis* 32(1) pp 11–29.

Cochrane WW (1958) *Farm Prices.* University of Minnesota Press, Minneapolis.

Comstock G (1987) *Is There a Moral Obligation to Save the Family Farm?* Iowa State University Press, Ames, IA.

Conway GR and Pretty JN (1991) *Unwelcome Harvest: Agriculture and Pollution.* Earthscan Publications Ltd, London.

Cramer HH (1967) 'Pflanzenschutz und Welternte'. *Bayer Pflanzenschutznachrichten* 20(1) p523.

Davies K, Wiles R and Campbell C (1996) *Pay to Spray: campaign contributions and pesticide legislation.* Environmental Working Group, Washington, DC.

Dinham B (1993) *The Pesticides Hazard.* The Pesticides Trust, London.

Dinham B (1996) 'Pesticide production in the South'. In *Growing Food Security: Challenging the link between pesticides and access to food*, pp 12–16. The Pesticides Trust, London.

Escalada MM and Heong KL (1993) 'Communication and implementation of change in crop protection'. In Ciba Foundation (1993) *Crop Protection and Sustainable Agriculture*, pp191–202. John Wiley & Sons, Chichester, UK.

EWG (1995) *Forbidden Fruit.* Environmental Working Group, Washington, DC.

Fagin D, Lavelle M and the Center for Public Integrity (1997) *Toxic Deception: how the chemical industry manipulates science, bends the law, and endangers your health.* Birch Lane Press, Secaucus, New Jersey.

Farah J (1994) 'Pesticide policies in developing countries: do they encourage excessive use?' World Bank Discussion Paper # 238, The World Bank, Washington, DC.

Farmworker Justice Fund, Inc (1990) Testimony for hearing record on 'Environmental toxins and children: Exploring the risks' in Congress of the US *Environmental toxins and children: Exploring the risks, Part II. Hearing held in Washington, DC before the Select Committee on Children, Youth, and Families.* ED 336 178. Government Printing Office, Washington, DC.

Feber R (1996) 'The effects of organic and conventional farming systems on the abundance of butterflies'. Report to WWF (UK): Project 95/93 – Plants and Butterflies: Organic Farms. Department of Zoology, Oxford.

Freemark K and Boutin C (1995) 'Impacts of agricultural herbicide use on terrestrial wildlife in temperate landscape: a review with special reference to North America'. *Agriculture, Ecosystems and Environment* 52, pp 67–91.

Gannon E (1992) *Environmental clean-up of fertilizer and agri-chemical dealer sites: 28 Iowa Case Studies.* Iowa Natural Heritage Foundation, Des Moines, IA.

Gardner BD (1995) *Plowing Ground in Washington: the political economy of US agriculture.* Pacific Research Institute for Public Policy, San Francisco, CA.

Garry VF, Schreinemachers D, Harkins ME and Griffiths J (1996) 'Pesticide appliers, biocides, and birth defects in rural Minnesota'. *Environmental Health Perspectives* 104, pp 394–399.

Ghersa CM, Roush ML, Radosevich SR and Cordray SM (1994) 'Coevolution of agroecosystems and weed management'. *BioScience* 44(2) pp 85–94.

Gianessi LP and Puffer CA (1993) 'Herbicide-resistant weeds may threaten wheat production in India'. *Resources* Spring 1993, pp 17–22.

Goodman D and Redclift M (1991) *Refashioning Nature: food, ecology and culture.* Routledge, London and New York.

Gore A (1994) Introduction to *Silent Spring.* Houghton Mifflin, New York.

Groth E (1996) Review of 'Our Children's Toxic Legacy'. *Science* 274, pp 61–62.

Hill SB and MacRae RJ (1995) 'Conceptual framework for the transition from conventional to sustainable agriculture'. *Journal of Sustainable Agriculture* 7(1), pp 81–87.

Howard A (1940) *An Agricultural Testament.* Oxford University Press, London.

Hynes, HP (1989) *The Recurring Silent Spring.* Pergamon Press, Elmsford, NY.

Ikerd J (1996) 'Ag industrialization: who should fund it?' *Agweek* 29 April, 1996, p5.

Iowa State University (1994) *1994 Iowa Farm and Rural Life Poll, 1994 Summary Report.* Iowa State University, IA.

Japan–MAFF (1995) *1995 Census of Agriculture.* Ministry of Agriculture, Fisheries and Food, Japan (30 November, 1995).

Kannan K, Tanabe S and Tatsukawa R (1995) 'Geographical distribution and accumulation features of organochlorine residues in fish in tropical Asia and Oceania'. *Environmental Science and Technology* 29, pp 2673–83.

Kloppenburg J Jr (1991) 'Social theory and the de/reconstruction of agricultural science: local knowledge for an alternative agriculture'. *Rural Sociology* 56(4) pp 519–48.

Kneen B (1993) *From Land to Mouth: understanding the food system.* NC Press, Toronto.

Knutson RD, Taylor CR, Penson JB and Smith EG (1990) *Economic Impacts of Reduced Chemical Use.* Knutson and Associates, College Station, Texas.

Lang T (1997) 'The public health impact of globalisation of food trade' in P S Shetty and K McPherson (eds) *Diet, Nutrition and Chronic Disease: Lessons from Contrasting Worlds.* John Wiley & Son, Somerset, New Jersey.

Levins R (1996) *Monitoring sustainable agriculture with conventional financial data.* The Land Stewardship Project, White Bear Lake, MN.

Lighthall DR (1996) 'Sustainable agriculture in the Corn Belt: production-side progress and demand-side constraints'. *American Journal of Alternative Agriculture* 11(4), pp 168–174.

Linstone HA (1994) 'New era – new challenge'. *Technology Forecasting and Social Change* 47, pp 1–20.

Lowe P (1992) 'Industrial agriculture and environmental regulation: a new agenda for rural sociology'. *Soliologia Ruralis* 32(1), pp 4–10.

Luck RF, van den Bosch R and Garcia R (1977) 'Chemical insect control – a troubled strategy'. *BioScience* 27, pp 606–11.

MacRae, RJ, Henning, J and Hill, SB (1993) 'Strategies to overcome barriers to the development of sustainable agriculture in Canada: the role of agribusiness'. *Journal of Agricultural and Environmental Ethics* 6 pp 21–51.

Moore CV and Villarejo D (1996) 'Pesticide cancellation and Kentucky windage'. *Choices* Third Quarter 1996, pp 36–38.

NRC (1996) *Ecologically Based Pest Management: New Solutions for a New Century.* Board on Agriculture, US National Research Council, National Academy Press, Washington, DC.

Norse D (1994) 'Multiple threats to regional food production: environment, economy, population?' *Food Policy* 19(2), pp 133–48.

OECD (1993) *Agricultural Policies, Markets and Trade: Monitoring and Outlook.* OECD, Paris.

Oerke E-C, Dehne H-W, Schönbeck F and Weber A (1994) *Crop Production and Crop Protection: Estimated Losses in Major Food and Cash Crops.* Elsevier, Amsterdam.

Oskam AJ (1995) 'The economics of pesticides: an overview of the issues' in *Policy Measures to Control Environmental Impacts from Agriculture* (Concerted Action AIR3-CT93–1164). Workshop on Pesticides, 24–27 August 1995. Wageningen Agricultural University, Department of Agricultural Economics and Policy.

PANOS (1996) *Briefing: Feast or Famine? Food Security in the New Millennium.* PANOS, London.

Paarlberg R (1996) Comments prepared for Yale University's Center for Environmental Law and Policy's 'Next Generation Project' expert symposium on agriculture, New Haven, CT, 7–8 March 1996.

Perkins JH (1982) *Insects, experts, and the insecticide crisis.* Plenum, New York.

Pfeffer MJ (1992) 'Labour and production barriers to the reduction of agricultural chemical inputs'. *Rural Sociology* 57(3) pp 347–62.

Pingali PL and Roger PA (1995) *Impact of Pesticides on Farmers' Health and the Rice Environment.* Kluwer Academic Press, Dordrecht, The Netherlands.

Pollack S (1990) 'Pesticide exposure and working conditions among migrant farmworkers' children in Western New York State'. Paper presented at the American Public Health Association Annual Meeting.

Ponting C (1993) *A Green History of the World.* Penguin, London.

Porterfield J, Rawlins S and Race L (1995) *Trends in pesticide use in US agriculture.* American Farm Bureau Public Policy Division, Illinois.

Pretty JN (1995) *Regenerating Agriculture: Policies and Practice for Sustainability and Self-Reliance.* Earthscan Publications Ltd, London; National Academy Press, Washington, DC; and Vikas Publishers, Bangalore.

Pretty, JN (1998) *The Living Land.* Earthscan Publications Ltd, London (in press).

Pretty JN, Thompson J and Hinchcliffe F (1996) *Sustainable agriculture: impacts on food production and challenges for food security.* Gatekeeper Series No 60, International Institute for Environment and Development, London.

Public Citizen (1996) *A million for your thoughts.* Public Citizen, Washington, DC.

Putnam JJ and Allshouse JE (1996) *Food consumption, prices, and expenditures, 1996.* Food and Consumer Economics Division, Economic Research Service, US Department of Agriculture, Washington, DC.

Repetto R and Baliga SS (1996) *Pesticides and the immune system: the public health risks.* World Resources Institute, Washington, DC.

Rola A (1989) *Pesticides, Health Risks and Farm Productivity: A Philippine Experience.* Agricultural Policy Research Program Monograph no 89–01. University of the Philippines, Los Baños.

Rola, AC and Pingali PL (1993) *Pesticides, Rice Productivity and Farmers; Health: An Economic Assessment.* International Rice Research Institute, the Philippines, and World Resources Institute, Washington, DC.

Rosenblum, G (1994) *On the Way to Market: Roadblocks to Reducing Pesticide Use on Produce.* Public Voice for Food and Health Policy, Washington, DC.

Rosset PM and Altieri MA (1995) 'Agroecology versus input substitution: a fundamental contradiction of sustainable agriculture'. Paper presented at The Politics of Sustainable Agriculture conference, 7–8 October 1995, University of Oregon, Eugene, OR.

Smith S (1992) 'Farming – it's declining in the US' *Choices*, First Quarter, pp 8–10.

Tenner E (1996) *Why Things Bite Back: technology and the revenge of unintended consequences.* Alfred A Knopf, NY.

Thrupp LA (1995) *Bittersweet Harvests for Global Supermarkets: Challenges in Latin America's Agricultural Export Boom.* World Resources Institute, Washington, DC.

Toppari J, Larsen JC, Christiansen P, Giwercman A, Grandjean P, Guillette LJ Jr, Jégou B, Jensen TK, Jouannet P, Keiding N, Leffers H, McLachlan JA, Meyer O, Müller J, Rajpert-De Meyts E, Scheike T, Sharpe R, Sumpter J, Skakkebæk NE (1996) 'Male reproductive health and environmental xenoestrogens'. *Environmental Health Perspectives* 104 (Suppl 4) pp 741–803.

Tovy J (1990) *Agricultural Ecology.* John Wiley & Sons, New York, NY.

Tye R (1997) 'Are water consumers bearing an unfair burden?' *Pesticides News* 35, p18.

Unnevehr LJ (1993) 'Suburban consumers and exurban farmers: the changing political economy of food policy'. *American Journal of Agricultural Economics* 75: pp 1140–44.

US Agricultural Retailers Association (1996) 'Food for Thought' Calendar. ARA, St Louis, MO.

USDA Economic Research Service (1994) 'Shifts in the US corn market'. *Agricultural Outlook* November pp 14–16.

US-EPA (1992) 'Regulatory impact analysis of Worker Protection Standard for agricultural pesticides'. Biological and Economic Analysis Division, Office of Pesticide Programs, United States Environmental Protection Agency, Washington, DC.

US-EPA (1996) *1994 Toxics Release Inventory: Public Data Release*. United States Environmental Protection Agency, Office of Pollution Prevention and Toxics, Washington, DC.

Vorley WT and Keeney DR (1995) 'Sustainable Pest Management and the Learning Organization'. Background paper prepared for the International Food Policy Research Institute (IFPRI) workshop 'Pest Management, Food Security, and the Environment: the Future to 2020', Washington, DC.

Waibel H (1986) *The Economics of Integrated Pest Control in Irrigated Rice*. Springer-Verlag, Berlin and New York.

WCED (1987) *Our Common Future*. World Commission on Environment and Development, Oxford University Press, Oxford and New York.

Wessel J (1983) *Trading the Future: farm exports and the concentration of economic power in our food economy*. Institute for Food and Development Policy, San Francisco, CA.

Wiles R, Cohen B, Campbell C and Elderkin S (1994) *Tap Water Blues: Herbicides in Drinking Water*. Environmental Working Group, Washington, DC.

Wilson J (1993) 'The BTO Birds and Organic Farming Project: 1 year on'. *New Farmer and Grower* spring 1993, pp 31–33.

Wirth K and Schima F (1994) *The Pesticide PACs: campaign contributions and pesticide policy*. Environmental Working Group, Washington, DC.

Wright A (1990) *The Death of Ramón González: The Modern Agricultural Dilemma*. University of Texas Press, Austin, TX.

3

The True Price of Pesticides

David Pearce and Robert Tinch

Introduction

The use of pesticides has its rationale in the extent to which they protect agricultural output. Put another way, the benefit of pesticide use is defined and measured by the value of the output that would otherwise be lost if pesticides were not used, or not used in current quantities. Pesticides are not free resources – a price is paid for their use. The question is, is the price paid by farmers the 'real' or 'true' price ? The answer to this question is in two parts.

Firstly, we know that in many parts of the world farmers do not pay the financial costs of production of pesticides. Prices are subsidized, often for very good motives, such as the need to keep agricultural productivity high in the face of food insecurity. But these motivations are also often misplaced, either because the goal of subsidization is not met, or because the subsidy encourages wasteful use of pesticides, harming the environment and human health. Secondly, even where farmers pay the financial cost of production, they may well not be paying the 'full' or 'social' cost of production. This is because pesticides have environmental and health impacts which market prices ignore. The market price of a pesticide does not, for example, include the impairment of health among pesticide users, nor does it include reductions in biological diversity that may result from pesticide applications, or any groundwater contamination, and so on. These 'external' costs are not paid because economic systems are usually set up in such a way that the interests of third parties who may suffer the incidental effects of production and consumption are not fully represented in the prevailing system of property rights. If, for example, pesticide manufacturers and users had always to account for their impacts on such third parties, we could expect the nature of production and use decisions to change. Less pesticide might be used because users would be aware that they would have to compensate potential sufferers. Or pesticides might be used more carefully. Without such property right arrangements there is said to be a 'missing', 'partial' or 'incomplete' market and this explains why governments often intervene in an effort to protect third parties, perhaps by banning especially dangerous pesticides or by regulating use rates. If pesticide use is associated neither with redefined property rights nor with regulations, then it is very likely to be excessive, and the external costs will be significant.

This chapter seeks to investigate this imbalance between the true and actual price of pesticides. In order to do this, it is necessary to look at the theoretical underpinnings of the argument that true cost and actual cost are likely to diverge.

THE BASIC ECONOMICS OF PESTICIDES

The concept of 'true' price

The resources which society uses to satisfy its desires are scarce: there is not enough land, labour, machinery, or raw materials to produce all the goods which people would like. This scarcity forces society to choose how resources are to be used and who is to receive the products. All societies solve this problem in some way. Wholly centrally planned economies, of which there are comparatively few now, determine the outcome by edict, through output targets, for example. In market economies, the problem is resolved through the price system, which determines the allocation of goods and services.

In a free market economy – one in which there are limited controls exercised by government on the forces of supply and demand – supply and demand will determine the price of the good. The willingness of producers to supply goods depends on the relationship between the market price and the cost of production. The cost of production, in turn, includes the costs of the land, labour, machinery, and so on used to manufacture and supply the product. This cost of production is also called the 'opportunity cost' of supply. Opportunity cost is the value of the output that could have been produced had the resources not been used for the product in question.

For an 'economically efficient' outcome, the benefit to society of producing any given good must be at least as great as its opportunity cost. If this were not true, more economic value could have been secured by switching the resources to an alternative use. This gives us the first insight into the *benefit-cost rule*: goods should be produced at the point where the benefit of producing the last unit is just equal to the opportunity cost of producing it (the forgone benefit from some other product). Going beyond this point means that the extra benefit is less than the extra cost. Keeping below this point means that there will exist some unexploited net gain from expanding output. This benefit-cost rule can be applied in the same way to inputs, such as pesticides. The extra benefit from one more unit of pesticide must be just equal to the forgone benefit by expanding pesticide use by one unit rather than using the resources as inputs to something else. The *demand* for pesticides actually reflects the benefits of their use, otherwise people would not wish to acquire them. Less obviously, the *supply* of pesticides reflects the opportunity cost of their production. They would not be supplied unless it was more 'profitable' to do so than to use the resources for some other purpose. This links the benefit-cost rule to the concepts of supply and demand.

The forces of supply and demand therefore allocate resources efficiently, leading to an allocation which maximizes the sum of net benefits to producers (profits: in other words the difference between price and cost of production) and net gains to consumers (known as 'consumer surplus', the difference between willingness to pay – or benefit – and price). The notion of profit is easy to understand. Consumer surplus is best understood as the gain that individuals get over and above the price they pay for something. Thus, if X is willing to pay $5 for something (he or she would have bought it at that price) but the ruling price in the market is $4, there is a surplus willingness to pay – the consumer surplus – of $1. Analysis would be easy, then, if the problem of the true price of pesticides could be analysed solely in the context of a perfectly functioning market in which the forces of supply and demand operate unhindered and in which there are no third party effects (no externalities). In that case, the true price would be the price ruling in the market and there would be no question of the 'wrong' allocation of resources to pesticides. There would also be no need for regulation of pesticide use. However, in practice, markets do not function this well. They fail to secure this maximization of net gains.

Market failure

The failure of markets to function in the perfect fashion indicated previously is due to various reasons:

- *Failure of competition* – the market should be competitive, ensuring production is efficient and preventing any one producer (or consumer) from making excessive profits. An example of this kind of failure is monopoly, where a single producer is able to increase the price, selling fewer goods, but making more profit on each one. This is good for the monopolist, but bad for society as a whole. Similarly, cartels of producers, or monopsonists (single consumers of a good or service) lead to inefficient outcomes. In monopolies, prices will tend to be too high and quantities too low compared to what would happen if competition existed.
- *Failure of information* – competition can only maintain efficiency if people know all their options. For example, if the purchaser of a good does not know that the same good is available at a lower price in the next shop, he/she will have no incentive to buy the least expensive (uncompetitive) product. Another form of information failure occurs where imperfect information about the future leads firms and regulators to be overly comfortable with the status quo, leading in turn to inefficiently low levels of investment in research and development. A highly relevant source of information failure for pesticides relates to the availability of agricultural extension services incorporating instruction on the safe use of pesticides, and courses provided by the multinational companies themselves. The same services could, of course, provide information on alternatives to pesticides, although multinational companies will have less incentive to encourage these. If farmers or farm workers are uninformed about the hazards of using pesticides, then they may use them in the wrong way, incurring health costs to themselves and others.
- *External costs* – prices reflect only those costs borne by the producer, or internal costs. Where the production activity creates costs which are borne by others, for example pollution damage, these external costs are not incorporated in the price. Then purchasers do not pay the full costs to society of producing the good, and therefore too much of the good is produced and sold. In the case of pesticides, the externalities are likely to be in the form of water contamination and species loss.

The main focus of this chapter is on the last aspect of market failure: the externalities. In the case of external costs, the inefficiency would be removed if producers and consumers considered social costs instead of simply private costs. Social costs are the sum of private (internal) costs and external costs. They represent the true cost to society of producing a good. We can write

$$\text{Social Cost} = \text{Private Cost} + \text{External Cost}$$

Note that costs, in the sense used here, take a wider definition than simply financial costs. A cost is the use of any scarce resource, including labour, natural resources, capital equipment, use of the environment, and so on. Money is used to measure costs, so that different physical costs can be compared. So long as the market operates properly, the prices used to value these physical inputs will be accurate reflections of their opportunity costs. But unless there is some form of regulation, external costs will not be reflected in market transactions and will therefore be given no price. But external costs can be significant. In the broader context, they include health damages from air pollution, global warming and, for example, damage to sensitive ecosystems from acid rain. In the context of pesticides, damages range from loss of biological diversity to groundwater contamination. The damage is real, and a cost is imposed on some or all members of society. But as there is no market in clean air or the global atmosphere, it is difficult to bring these costs into a direct relationship with private costs.

Solutions to market failure

One possible solution to the externality problem would be to create a market in the environmental good. For any market to operate, there must be ownership of the good traded. Ownership of the right to pollute, either with or without limits, could be given to firms and individuals who engage in polluting processes. Then those suffering damage could buy some of the right to pollute from the polluting firms, paying them to emit less pollution. Alternatively, damage sufferers might be given ownership of a clean environment. Then firms and individuals wishing to create pollution would have to buy the right to pollute from those who suffer from pollution. In either case, both sides would agree to a trade if, on balance, they benefited from it.

The idea of owning the environment may seem strange, but in effect it is unavoidable: ownership occurs by law or by default. Whenever a firm emits pollution, it is acting with de facto ownership of the environment. When government decides to set limits on pollution, it is reclaiming some of that ownership right for others. The polluter pays principle can be interpreted as saying that the ownership of all environmental goods should lie with society as a whole, and that individuals who want to make use of the environment by polluting it should pay for the cost. In practice, while cases of trading between polluters and sufferers exist, they are not many. The main reason for this is that there is usually a large number of sufferers, rather than just one. Often, the number of polluters is also large: this is the case where many farmers apply pesticides in the same water catchment area, for example. Under these circumstances, the procedure of bargaining, and of deciding who should pay what to whom, becomes very complicated and therefore costly.

A second alternative is for government to act to secure an efficient outcome. This could be achieved, for example, through setting environmental standards or through environmental taxes. The idea of environmental taxes is that the value of external costs is added to the price of the good – then, the price reflects all the costs and the outcome will be efficient. Economists tend to prefer taxes to standards where the choice exists. This is because they offer more flexibility. Often, it will cost different firms different amounts to cut pollution. Under a tax, the firms will take into account the costs of reducing emissions and compare these with the tax. Firms will reduce pollution where it costs them less to do so than it would to pay the tax. So firms which find it cheaper to reduce pollution will cut emissions more than firms finding abatement measures more expensive. Pollution abatement will occur where it is least expensive. Under a regulatory standard, this flexibility is not available. The same extent of pollution reduction can be achieved more cheaply – for society – under a tax system.

Environmental tax revenues could be used to reduce other forms of taxation, or to increase government spending on, say, the environment or health or education. This is often termed the 'double dividend' of environmental taxes – they achieve pollution reduction at minimum cost, while simultaneously providing funds for the creation of other benefits. The money could even be returned to industry, for instance through reducing other taxes, especially taxes on income and effort which themselves distort economic efficiency, or grants for less polluting technology. So long as the money was spread evenly among firms, or went to less polluting firms only, the incentive provided by the tax to cut pollution would remain.

In many cases of environmental pollution, it is not feasible to measure directly the damage caused at each moment in time. Where emissions can be measured, they can be used instead as a proxy for damage caused, provided the relationship between emissions and damages is judged to be a proportional one. Similarly, by estimating the relationship between pesticide inputs and damage, a tax on pesticide use can be devised which is equivalent to a tax on marginal damage. In addition, taxing inputs, such as pesticides, simplifies the issue where there is more than one sort of damage to be considered – for example in pesticide use, where there is contamination of water and direct effects on animal life within the treated area. The drawback with taxing inputs instead of damage is that the linkage between inputs and damage is usually not perfect. For example, pesticide contamination of water resulting from a given pesticide input will vary with the weather, the soil structure and the distance from the field to a watercourse. Just as important, pesticide damage may be separated from the application of the pesticide by long time lags: damage in one period may be related to 'emissions' years, perhaps decades previously. Not controlling these factors in the setting of a tax will lead to a less efficient outcome than might be achieved in an ideal world. However, designing and operating a more perfect system, which incorporated all these differences, or which measured damage directly, would be much more expensive than a simple tax on pesticide use. Realistically, the inefficiency introduced by relying on a simple pesticide tax is likely to be much less than either the inefficiency of allowing unrestricted use of pesticides, or the costs of administering a more complicated system which took more factors into account.

Nevertheless, any pesticide tax should attempt to take into account differences among different pesticides. A uniform tax applied to all pesticides would only create incentives for reduced pesticide use, whereas a tax which varied according to the damage potential of different pesticides would provide incentives to switch away from the most damaging chemicals towards the less damaging, as well as providing incentives to reduce overall use of pesticides. A tax system which takes into consideration the great diversity of pesticides will result in a much more efficient outcome than a uniform tax which does not. For an attempt to do this, see the analysis in Chapter 4.

The other main dimension of solving market failure is to improve the flow of information about pesticide risks. In a review of the price and non-price factors encouraging excessive pesticide use in developing countries, Farah (1994) reports that the most important non-price factor is the absence of adequate information for policy-makers or implementors and farmers on adequate and simple procedures to use pesticides, to define pests and crop loss, and alternatives to chemical pesticides. Related to this, both governments and external donors emphasize the information regarding chemical control methods while neglecting alternative approaches. In a study of the Philippines, Warburton et al (1995) found that farmers understood that pesticides are hazardous, but tended not to differentiate between pesticides according to degrees of hazard as defined by the World Health Organization (WHO). While it seems reasonable to suppose that hazards are correlated with degree of pest reduction effectiveness, in fact farmers tended to use odour as the main indicator of hazard. Farmers and labourers took protective measures but few seemed to understand that pesticides could be absorbed through the skin, and the measures taken were actually inadequate. Thrupp (1990) shows that banana growers in Costa Rica used excessive quantities of copper-based fungicides, to an extent that the relevant soil was contaminated and rendered unfit for use. Thrupp suggests that ignorance and poor flows of information were significant factors in the problem. The relevance of information failure to pesticide control policy is exemplified by the fact that it may be the flow of information that has to be controlled and improved rather than, say, the price of pesticides.

Government failure

It is tempting to think that government intervention will solve market failure, but when governments intervene we often find another form of inefficiency: government failure. Governments may produce less efficient outcomes than if they had allowed the market to allocate resources. Inefficient interventions may include subsidies, price controls, physical output targets, exchange controls, ownership controls, and so on. Even when intervention is justified, as it is in many contexts, the government should secure least-cost interventions – that is, use instruments of control that are not unnecessarily expensive.

Government failure plays a particularly important role in pesticide use, which is widely subsidized in the developing world. Keeping the price of pesticides artificially low has a number of effects:

- It encourages excess applications (applications over and above the recommended levels for crop protection). Subsidies also encourage pesticide *misuse*. Examples of misuse include use on crops for which a pesticide is not approved, failure to observe harvest intervals, careless disposal of unused pesticides and containers. Misuse and overuse are quite distinct concepts, and the policies required to address them will generally be different.
- It biases the choice of pest control technology towards pesticides rather than towards integrated pest management (IPM).
- It absorbs valuable government income which could otherwise be used for alternative developments, including IPM and other sustainable agriculture practices.

The average rates of subsidy in several developing countries for the mid 1980s are shown in Table 3.1. Most of these subsidies were put into effect during the green revolution in an attempt to encourage small farmers to adopt the pesticide technology that was needed to grow high-yield varieties of crops. Since then a number of countries have sought to reduce subsidy levels. Farah (1994) assembles the evidence as far as it is available for recent experience. The picture is clearly uncertain, but the general impres-

Table 3.1 *Pesticide subsidies in the developing world in the mid 1980s*

Country	*Subsidy (% of retail cost)*	*Total value ($ million)*	*Value per capita ($)*
China	19	285	0.3
Colombia	44	69	2.5
Ecuador	41	14	1.7
Egypt	83	207	4.7
Ghana	67	20	1.7
Honduras	29	12	3.0
Indonesia	82	148	0.8
Senegal	89	4	0.7

Source: Repetto (1989)

sion is that subsidies to pesticide use remain widespread in some countries but that other countries have at least begun a process of their gradual removal.

'Optimal' pesticide use

The costs and the benefits of any activity do not necessarily increase in a one-to-one relationship with the activity level. For example, a consumer gains a certain benefit from eating an apple. The benefit from a second apple may be similar or a little less. A third or fourth apple would bring less benefit than the first. At some level, no benefit would be derived from eating a further apple; beyond this, the consumer is likely to become so heartily sick of apples that eating another would be a cost, rather than a benefit. Economists use the concept of the margin to determine the best or optimal amount of consumption of a product, and, similarly, the optimal level of an input, such as a pesticide. The marginal benefit of an activity is the *addition* to benefits brought about by a one unit increase in the activity over its current level (for example, eating one more apple).

When the level of any activity increases, benefits increase by the amount of marginal benefit (hereafter, MB), while costs increase by the amount of marginal cost (MC). So, on balance, the benefit to society changes by +(MB – MC). Clearly, so long as (MB–MC) is positive, increases in the activity bring net benefits. When it is negative, reductions in the activity bring net benefits (because reducing an activity loses MB but saves MC). But when it is zero, which is when marginal benefits and marginal costs are equal, there is no benefit in changing the activity level in either direction. *Net* benefits are then maximized. Usually, marginal costs and marginal benefits vary with the activity level. For example, the marginal benefit of a unit of pollution reduction (that is, the damage avoided) is likely to be greater when pollution levels are high than at lower levels. Conversely, the marginal cost to a polluter of making a reduction is likely to be lower for the first few units than when emissions have already been cut substantially.

This reasoning tells us that an efficient outcome must be one in which marginal costs and benefits are equal. Otherwise, more benefit could be derived if the activity level was changed. A well functioning market creates such an outcome: individuals consider their marginal benefit for each good, choosing consumption levels to set this equal to the price. So long as the price of each good reflects the marginal costs of its production, this ensures that MC = MB = price, and the outcome is efficient. The same analysis can be applied to pesticide use. The optimal level of pesticide use is where the marginal cost of applying one more unit of pesticide is just equal to the marginal benefit. But the *perspective* now matters. From the point of view of the farmer, the relevant equation compares the marginal cost of the pesticide, which will be equivalent to the price of a unit of pesticide, and the marginal benefit: the value of the extra output of crops protected by using the pesticide. For the farmer, then, the equation becomes:

Pesticide Price = Extra Value of Crop Protected

In this case, the pesticide price is the private cost to the farmer. From the standpoint of society, however, the equation is incomplete. If the pesticides impose external costs on third parties, then the true cost of pesticide use is not the pesticide price, but the price plus the value of the extra damage done to third parties. This combination of the two costs is the concept of social cost introduced previously. The equation for optimal use of the pesticide now changes to:

Pesticide Price + Value of Extra Damage Done to Third Parties =
Extra Value of Crop Protected

The economics illustrated: A diagrammatic analysis

The previous sections have described the elements of the economics of pesticide use in literal terms. This section introduces the same theorems but in diagrammatic form. More sophisticated approaches still can be found in Carlson and Wetzstein (1993).

Figure 3.1 (top half) shows the basic relationship between pesticide applications and crop output protected. The function shown has a curvilinear shape, reflecting the probability that, as pesticide applications increase, output protected will increase. However, as shown, the extra protection declines as more and more pesticides are applied, reflecting diminishing *marginal* returns in crop value protected (the slope of the curve shows the marginal returns). Eventually, very large applications of pesticide may actually damage total output, and the curve begins to decline. The curve shown in the top half of Figure 3.1 is a *pesticide production function.*

The previous discussion shows that to move towards an analysis of optimal application rates we need to work in marginal units. The lower half of Figure 3.1 is actually the slope of the production function. What it measures then is the ratio:

$$\frac{\Delta \text{ crop output protected}}{\Delta \text{ pesticide application}}$$

where Δ simply means ‘change in’ or ‘marginal’. Now, if we multiply the output protected by the price of the output (crop), the top line becomes the *value* of the extra output protected. This ratio is known as the *value of the marginal product* of pesticide application. It is particularly important because it is, in fact, the farmer’s demand curve for pesticides. And this is another name for the *marginal benefit function.* This is most readily seen in the lower half of Figure 3.1 by comparing the curve against price on the vertical axis. At pesticide price P1, the farmer will wish to buy pesticide quantity A1. This is because buying A1 maximizes the net gains to the farmer. If he buys more than A1, say A2, he will be getting less benefit from the application of A2–A1 than the price paid for the pesticide. In terms of the diagram, he would have a net loss on A2–A1 of the shaded area. If he buys less than A1 he will be missing the opportunity for further

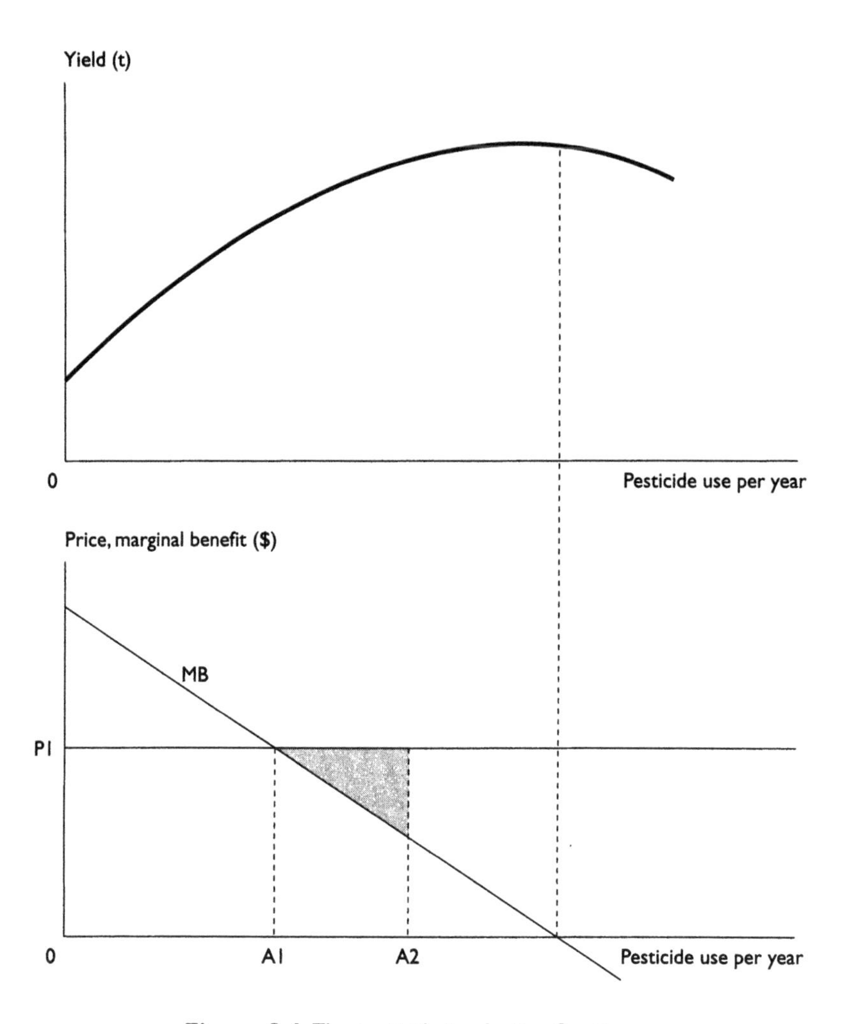

Figure 3.1 *The pesticide production function*

net gain. A1 is therefore the 'privately optimal' level of pesticide use. Of course, the diagram is simplistic since we know that choosing optimal pesticide applications involves many other issues, including those of timing, minimum pest density and so on (see Carlson and Wetsztein, 1993). But the diagram conveys the essence of the farmer's own cost-benefit decision.

Figure 3.2 introduces the idea of subsidies. For the individual farmer the price of pesticides is given. It is determined in the market place. Provided the individual farmer is not powerful enough to influence the market price, price can be shown as a horizontal line, just as it is in Figure 3.1. A subsidy that is a constant percentage of the price would then be shown as a horizontal line but lying below the original price. Thus, if P1 is the price without subsidy, P2 may be the price with a subsidy. Of course, subsidies may work in complex ways and Figure 3.2 simply indicates a very basic idea of a subsidy as a constant percentage of price. Notice that the assumption in economic models of this kind is that it is the farmer who makes pesticide decisions. In practice, many agents

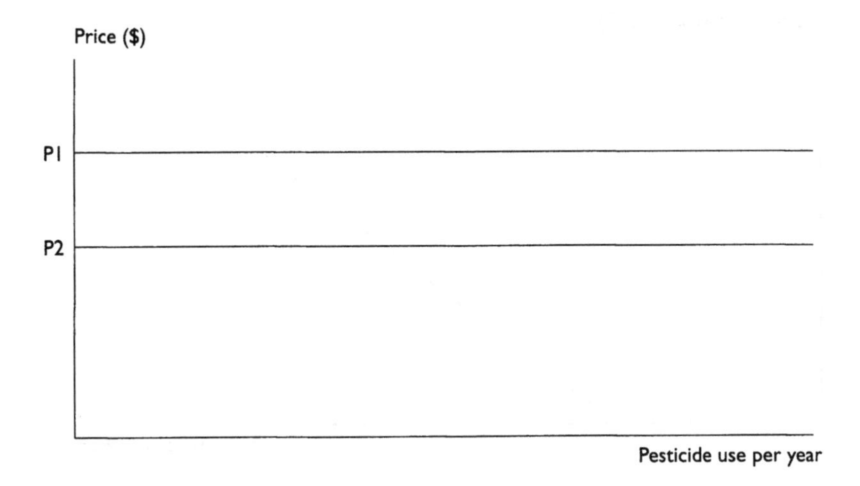

Figure 3.2 *Subsidies and the price of pesticides*

are involved, as Zilberman et al (1994) note. Once the focus shifts to markets as a whole, the pesticide price curve will tend to slope upwards to reflect rising marginal costs of production. Or it may slope downwards if there are significant economies of scale in production. We retain the constant function for simplicity of exposition.

Figure 3.3 now introduces the *marginal external cost function* (MEC): the monetary value of the environmental damages done by extra applications of pesticides. Note that it is in monetary terms, and this raises the whole issue of the monetary valuation of externalities. This will be discussed in Chapter 4. For simplicity we have assumed the MEC function is a straight line and upward sloping. Other shapes are possible. As shown here, MEC does not start at the origin. This is because very low application rates are assumed not to produce any harm to the environment. If this assumption does not hold and low doses do produce harm, then the MEC curve would begin close to, or at, the origin. The correct assumption is very much a matter for the toxicologist and the analysis of *dose-response functions.*

Figure 3.4 brings all the pieces together. Consider first the situation from the perspective of the farmer. If there are no subsidies, he will adopt quantity $A\pi$. The rationale for this was given in the analysis of Figure 3.1: it is the application rate that maximizes the net benefits to the farmer. Now consider what happens if there are subsidies. The lower price line is now relevant and the farmer will purchase quantity AS of the pesticide. The rationale is straightforward: the cost to the farmer is now less than it was without the subsidy and so he will maximize his own net gains by increasing the purchase of pesticides.

Finally, consider the socially optimal application rate. From society's point of view it is important to include the external costs in the measure of true costs of pesticide application. Hence the relevant cost curve is now MSC. Society maximizes its net benefits from pesticide applications by setting MSC = MB, in other words by applying A* unit of pesticides. Note that the socially efficient level of pesticide use is associated with positive externality levels: in general, it is efficient to have some externality (called the *optimal externality*). This is because, on the one hand, it would be extremely expensive to eliminate environmental damage completely, while on the other hand, the

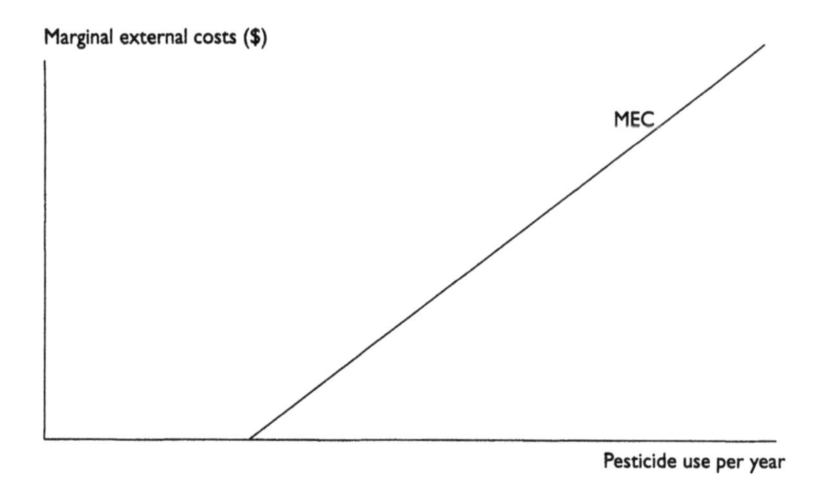

Figure 3.3 *Marginal external costs of pesticide use*

damage caused by low levels of pollution may not be particularly great. Although we would like to have zero pollution, we have to weigh this goal against the things we would have to give up to achieve it.

Since marginal curves are the slopes (derivatives) of total curves, areas under marginal curves equal the totals in question. For example, in Figure 3.4 the optimal externality – the optimal level of damage done to the environment – is given by the dark shaded area. Similarly, if we wish to find the net benefits to society, these are maximized at A* and are equal to the dotted area in Figure 4. In turn, this dotted area is equal to the area under the MB curve, *minus* the costs of production (the area below the price line), *minus* the optimal externality.

We can use the same analysis to illustrate the idea of a pesticide tax discussed in

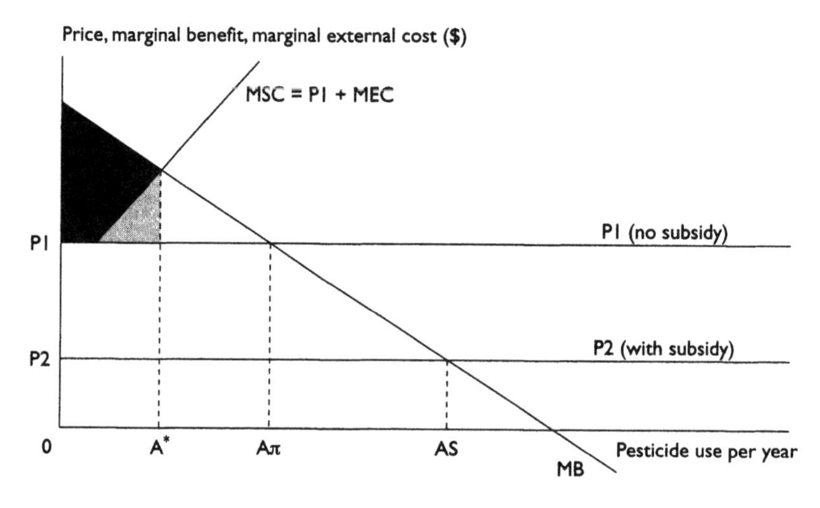

Figure 3.4 *Socially optimal pesticide use (subsidies)*

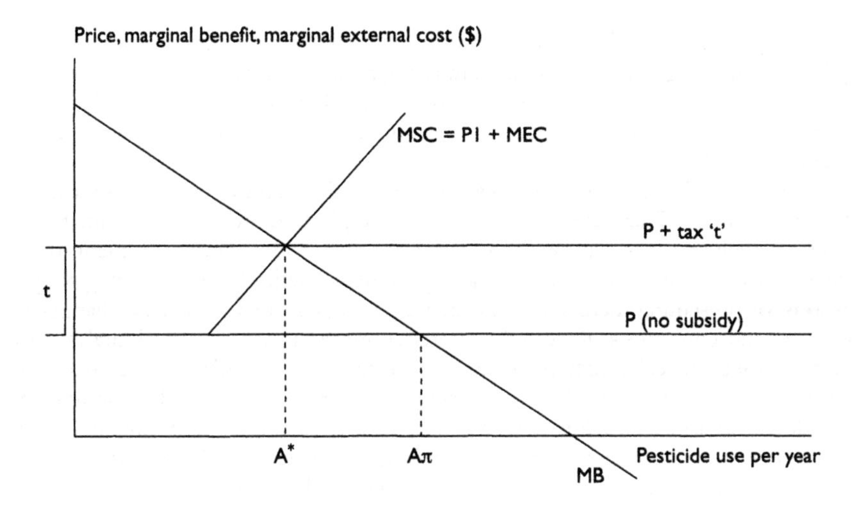

Figure 3.5 *Internalizing externalities of pesticide use*

outline previously. Figure 3.5 repeats the essence of Figure 3.4 but omits the subsidy issue. As before, Aπ is the privately profitable application rate, but A* is the socially desirable level. Since the farmer is unlikely to reduce application rates from Aπ to A* voluntarily, he needs to be regulated in some way. One approach would simply be to issue a law requiring the application rate to be A*. If pesticide use guidance was based on the principles of economics this would be one solution to the problem. In practice, they are more likely to be guided by the consideration of what is best for the farmer – they may come closer to Aπ than A*. An alternative approach to this standard setting solution, then, is to impose a tax on the pesticides themselves. Ideally, the tax would be equal to the distance 't' in Figure 3.5. To see this, simply add t to the price (no subsidy) of the pesticide to obtain the price (P + tax 't') line. Since the farmer will now regard this line as his cost of pesticide use, he will automatically opt for A*. Such taxes are known as *Pigovian taxes* after their originator, the Cambridge economist Arthur Pigou. In practice, estimating 't' is extremely difficult. This is because 't' is determined by the marginal external costs at the optimum. It is not, for example, equal to MEC at Aπ. This means we need knowledge of the whole MEC function and it is generally rare for this to be known. Nonetheless, the idea of a pesticide tax remains valuable and, provided we know the direction of control, such taxes are a powerful means of ensuring reduced pesticide use.

Environmental valuation

Figure 3.5 showed that in order to set a tax at the correct (socially efficient) level, we need to know what the costs of environmental pollution are. This is unlike the situation in which the polluter and sufferer trade directly; then, these parties decide for themselves what costs they face and act accordingly. To set a tax equal to marginal external cost, a policy-maker must estimate the marginal benefits other individuals would gain from reductions in pollution. As noted above, it is unlikely in practice that we could identify the socially optimal level of pesticide use with any great precision. But valuation is still important because we need to know if the externalities associated with pesticide use are significant or not. Listing the impacts is essential, but if there are different types of externality – health, biodiversity loss, water contamination – it will help if they can be aggregated in some way. Using money as a measuring rod is one way of aggregating such impacts and this is the procedure that has been used in at least one of the major attempts to estimate the size of the total external cost of pesticide use (Pimentel et al, 1992).

Environmental valuation techniques attempt to estimate the cost of damage caused to the environment. There are two main approaches, both of which seek estimates of individuals' willingness to pay (WTP) to secure an environmental improvement, or to avoid an environmental deterioration. Stated preference methods involve asking people questions, or presenting them with hypothetical choices, related to specific changes in environmental conditions. Revealed preference methods are statistical analyses of peoples' choices in real situations where they have options relating to the quality of their environment. An example of the first is *contingent valuation*, in which individual respondents are asked questions which reveal their WTP for an environmental good (for instance, a certain improvement in drinking water quality). An example of the second is *hedonic price estimation*, in which differences in the prices of similar marketed goods (for example, houses) are analysed to reveal the implicit price of an environmental good (local water quality), from which WTP for better water quality can be estimated.

Often, rather than attempt directly to estimate costs by one of the above routes, it is preferable to estimate 'dose-response' functions between emission levels and physical measures of impacts. Dose-response estimation is not an alternative to valuation. Rather, it is one stage of it, allowing complicated relationships to be broken down into more manageable parts. For example, we can use scientific knowledge to determine the effects of emissions in terms of various health risks, direct losses of animal life, indirect ecological effects, and so on. We can then attempt to value these individual effects. Estimating the physical effects or risks and subsequently multiplying by 'unit prices' (the willingness to pay for each unit of a physical effect) is often easier (and cheaper) than trying to conduct an original valuation study for each case. It may also give better results, avoiding problems associated with cases in which individuals' perceptions of risks differ from the actual risks to which they are exposed (see Chapter 4).

This is no more than the briefest of introductions to valuation concepts and methodologies. Full details are available in Freeman (1993) and Pearce, Whittington and Georgiou (1994).

Summary so far

Pesticides generate a benefit in the form of crop damage avoided. Pesticides also have costs. The direct or private costs are given by the price paid for pesticides in the market place. But there are indirect costs – the external costs – which must also be accounted for. The aim should be to internalize these costs, making private farmers take them into consideration, perhaps by a tax or by some other means. Taxes should be proportional to the damage done, and hence there is a need to measure the external costs of pesticide use. Farmer health costs may or may not be externalities. This depends on the state of farmers' information about the effects of pesticides. The more informed the farmer, the less likely it is that any health damage is external.

From Theory to Practice: The Benefits of Pesticide Use

Use of pesticides

The world pesticide market is valued at around $31–37 billion each year. The US alone is spending $11.3 billion (1995 figures) (US EPA, 1997). The actual amount of pesticides used is measured as the mass of active ingredient (AI) applied, because while a pesticide often contains inert substances, it is the active ingredient which confers pesticidal properties. However, inert ingredients can also have toxic effects on the environment. Summing active ingredients by mass is imperfect, since active ingredients vary widely in potency and in damage potential. For example, one metric tonne of pyrethroid is as potent as three to five metric tonnes of carbamate or organophosphate, or 10 to 30 metric tonnes of DDT. So a more meaningful measure of aggregate pesticide use would be to weight different active ingredients according to properties such as toxicity and persistence, and then sum the weighted totals.

There are two main difficulties to be overcome. Firstly, there exist many scientific data gaps, so that it is not known for all pesticides what the different damage effects are on wildlife, human health, and so on. And secondly, even if full information on these effects existed for all pesticides, some meaningful way of combining the different physical measures of damage would have to be found. In effect, this means that there is no way round the fact that any measure of aggregate pesticide usage which aims to take account of the different damage potentials of different pesticides must first successfully address the issue of how to value each individual form of damage.

The benefits of pesticide use

It is important to have some idea of the benefits of pesticide use for several reasons. Firstly, in order to calculate a pesticide tax close to the optimum of Figure 3.5, information on the marginal damage costs of pesticide use is of primary importance. A regulator could find the optimal tax by an iterative process of adjusting the tax rate until an equilibrium point was found at which the tax was equal to marginal damage. If marginal benefits are known, it allows the optimal tax to be set straight away, rather than being discovered through a sequence of adjustments.

A second reason for estimating the marginal benefits of pesticides is that the information is likely to be of great use to farmers. Straightforward economic theory assumes that individuals know what the benefits of their actions are. Where these are uncertain, they know the probabilities attached to the different possible benefits. But in reality, farmers may not be sure of the marginal effects of pesticide applications under any given condition. Setting the optimal tax will only result in an efficient outcome if farmers react according to their true marginal benefits, and so any relevant information with which they can be provided will aid the quest for efficiency. Thirdly, if the policy instrument chosen is not a tax, but is instead regulatory in nature, the regulator will need to know both the marginal benefits and the marginal damage costs of pesticide use in order to set efficient standards. In effect, regulatory instruments remove choice from the farmers, who are no longer able to vary their use of pesticides according to their private costs and benefits. Instead, the regulator chooses the level of inputs, and in order to make this choice efficiently, must know both the costs and benefits consequent upon it. Finally, a better understanding of the benefits of pesticides is required if the debate over the costs and benefits of pesticide use is to be informed. Focusing solely on the environmental costs is unhelpful if it neglects the very purpose of pesticide application – protecting food output. Similarly, focusing solely on farmer benefits will neglect the environmental dimension.

Whichever instrument is chosen, research into both the environmental damage costs and the crop protection benefits of pesticide use is required. Each has been the subject of various studies, but few of these have taken the requisite approach. For an efficient policy, it is necessary both to focus on the marginal effects of pesticide use, and to consider these effects for each different pesticide individually. In practice, most studies look not at a gradual or even partial reduction in pesticide use, but analyse instead the results of sudden changes in which pesticide inputs are reduced by 50 per cent or 100 per cent. The typical approach therefore estimates costs and benefits which are more total than marginal. In addition, the reductions postulated tend to be across the board cuts, allowing for no differentiation among pesticides in terms of their effects, nor for switching between different chemicals. Compared with more realistic changes, this procedure is likely to understate the external costs of pesticide use, because these costs are likely to fall less than proportionately with the extent of the reduction. Similarly, it is likely to overstate the benefits of pesticide use, because the effects on crop production of reducing pesticide use are likely to increase more than proportionately with the extent of the reduction. For example, the external cost saving from completely eliminating pesticides will be less than ten times, and the yield loss more than ten times, the cost saving and yield loss from a 10 per cent reduction in pesticides. The following are some studies attempting to measure the benefits of pesticide use.

Macro (whole economy) benefit studies

The benefits of pesticide use should be the increased output and better quality of food. Pimentel et al (1992) report a benefit-cost ratio for pesticides of 4:1 in terms of output and pesticide expenditure. Zilberman et al (1991) report a consensus estimate from surveying the literature that suggests benefit-cost ratios of 3 to 6.5. These results tend to come from mixes of expert judgement and some limited econometric studies which attempt to estimate the marginal product of pesticides (see Figure 3.1). The early work of Headley (1968), for example, suggests marginal products of four to six relating to the 1960s. Fischer's (1970) work in Canada suggests 2 to 13 for the 1970s, and Campbell's (1978) work for British Columbia suggests 12 to 13 for the late 1970s. Carlson's (1977) work for insecticides in various parts of the US generally suggests a wide range of marginal products of 1 to 50 for the 1960s. These studies appear to be the basis for the widely quoted 'average' estimate of around four for the US. Effectively, then, the total benefit of pesticide use is some four times the cost. Thus, for the US, the total cost of pesticide use in 1991 was $4.1 billion. Using the ratio of 4:1 would suggest, roughly, that crop damage avoided was $16.4 billion. One cannot conclude from this that the external costs would have to be of the order of (16.4–4.1) $12.3 billion for further controls to be exercized on pesticide use. To see this, imagine that the status quo is, in fact, optimal from the farmer's point of view. Then, we would be at $A\pi$ in Figure 3.5. The presence of positive external costs would then dictate a reduction in pesticide use. However, benefit assessment is itself complicated and should allow for quantity, quality and price effects. A proper approach here would involve the assessment of consumer and producer surplus changes in a general equilibrium setting to allow for changes in benefits and costs to crop farmers, livestock producers and consumers. Few of the studies attempting to estimate the benefits appear to adopt such a methodology.

Buttel (1993) provides a review of some of the 'benefit' studies. Generally, these look at the effects of removing pesticides on some or all of the following: farm incomes, prices, costs, exports, national income. Table 3.2 indicates the general findings based on Buttel (1993) and other references. Lee (1992) summarizes the results of these macro studies as:

- a general decline in yields;
- decrease in aggregate output;
- increased farm incomes; and
- consumer price increases.

Expressed more formally, farmers may actually gain from a switch to more organic approaches (provided most farmers make the switch), while consumers would lose. Indeed, one of the problems with the mounting number of studies on the impacts of technology switches is their failure to measure welfare losses and gains to the relevant groups (see Chapter 2).

Table 3.3 reports on studies which have estimated the costs of pesticide bans. Zilberman et al (1991) suggest that bans impose significant costs on producers and consumers when available substitutes are few, and when demand elasticities for the affected product are low, as would be expected. Bans in one region may also result in production shifting to other regions.

There are two clear camps on the issue of the benefits of reduced usage: those who argue that farm incomes would at least not fall, and environmental quality would increase, for a significantly reduced usage of pesticides (generally 50 to 100 per cent); and those who argue that such reductions would significantly increase costs and prices and have no effect on environmental quality, or even a deleterious impact. The latter school is in turn divided between those who believe that farm incomes would fall with reduced pesticide use, and those who believe that farm incomes may increase due to the relevant elasticities of demand. The debate between the various schools is fairly strong. Gianessi (1991), for example, mounts a fierce attack on the work of Pimentel et al (1991c) for:

- using 1970s pesticide use data to draw conclusions about impacts in the 1990s;
- extrapolating findings in one US state to other states, even though pest incidence varies by management practice across states;
- misuse of studies;
- specification of alternative technologies that are not practical;
- recommendation of alternative technologies that would reduce yields and some of which would harm the environment;
- recommendation of technologies that are widely used on the assumption that the scope for their further extension is substantial.

Clearly, there is room for significant debate over the benefits of pesticides relative to the chosen baseline of alternative pest management technologies.

Rather than look at dramatic changes in pesticide use, an alternative approach is one in which trends continued are modified by an assumed reduction in pesticides over and above any reductions already accounted for in the established trends. The differences in benefits to the various affected groups then define the benefits of pesticide use. Some of the literature argues that, even at this level, the benefits of pesticide use may not be significant because sustainable agriculture alternatives can compete with existing high-input agriculture in many instances (Faeth and Westra, 1993; Pretty, 1993, 1995). As noted above, if it is true that output can be at least maintained with a sustain-

Table 3.2 *Studies of effects of reduced pesticide use*

Study	*Output Effect*	*Exports*	*Prices/Costs*	*National Income*	*Crop Income*	*Livestock Income*	*Environment*
Olsen,1982, US	50% decline for wheat, corn, soya	negative	increase	negative	positive (due to demand elasticity)		
Knutson, 1990, USA	corn –32%; soya –37%; wheat & peanuts –78%	negative	100–200% increase in costs; $228 increase on annual household bills	negative	positive	negative (due to rise in food grain prices)	output decline stimulates extensive use of land
Ayer et al, 1990, US	criticizes Knutson study						
NAS, 1989, US	criticized by CAST, 1990				Y-chem = Y+chem farm income need not fall		
NRDC, 1991, US							
Pimentel, 1991, USA (50% pesticide reduction)	no decline in yields		0.6% increase in food prices				substantial environmental benefits in some areas
GRC, 1990, US	27 to 80% declines for cotton, soya, rice, wheat and corn		45% increase in Consumer Price Index for food				
Giannessi, 1991 (critique of Pimental, 1991)	yields would decline		costs would rise				
Gren, 1994, Sweden					fall by 8–28% for a 50% reduction in pesticides		
Tweeten and Helmers, 1990, US	11–22% decline for corn, grain, sorghum and soya		up to 87% increase in product prices		net returns on variables costs rise to 58%		

Source: Buttel (1993), Gianessi (1991)

Notes: Y-chem = income of farmers not using chemicals; Y +chem = income of farmers using chemicals

Table 3.3 *Effects of pesticide bans and controls*

Regulation	*Effect*
Ban on 1,3-D (soil fumigant) in California	Stanford Research Institute: $25 million extra expenditure by growers on more expensive substitutes in 1991; plus $75 million crop losses from nematode damage.
Ban on herbicides for lettuce in Florida, 1982 onwards	Increased production costs of $2 million per annum for lettuce growers 1989 defoliation from leafspot
Withdrawal of captafol in Michigan tart cherry orchards, 1987	Alternatives used but 1990 cherry production fell 12%, prices up 20%.
Restriction on use of atrazine herbicide in Wisconsin	Substitute herbicides now cost $7 to $11 per acre more.
Withdrawal of Alar	1987–1992, 30,000 Stayman apple trees taken out in Pennsylvania – previous production was heavily dependent on use of Alar to prevent cracking of apples.
1983 EPA ban on EDB (soil fumigant)	South-eastern soybean growers had to switch from $12 per acre EDB to substitutes at $50–$60 acre; 4.5 million acres of soybean production in Georgia, Alabama, and South Carolina stopped.
1986 Dinoseb herbicide ban	Predicted 20% peanut yield decline avoided by substituting paraquat.
EPA ban on DBCP fumigant to control nematodes	Citrus yields increased due to availability of fenamiphos and aldicarb.

Source: Gianessi (1993)

able agriculture system (if benefits are constant), and the combined private and external costs of sustainable agriculture are less than the private and external costs of intensive agriculture, then there would be no need for a benefit-cost evaluation: sustainable agriculture would simply be better.

The absence of a trade-off between reduced pesticides and food benefits also arises in the context where the pesticide problem is mainly one of overuse, as is certainly the case in many developing country contexts (see below). Pesticide usage may well exceed recommended doses, so that excess applications yield no additional benefits, only costs in the form of health and environmental damages. Obviously, reductions in 'excess' use must then result in all-round gains to farm incomes (reduced pesticide expenditure without reduced pesticide benefits) and to the environment and health. Such gains are akin to reductions in 'X-inefficiency' in standard industrial economics.

Buttel (1993) notes that 'prochemical researchers have stressed the simulation of the socioeconomic impacts of 100 per cent bans on chemicals..., while pro-sustainability researchers have stressed ... fractional reductions (typically 10 to 50 per cent) in chemical use'. The reasons are clear enough: a total ban on chemicals would be very disruptive and would preclude the use of pesticides even where the benefits they brought were significant. So the losses estimated will be high, helping the 'prochemical' case. Conversely, small changes in chemical use could often be accommodated with minimum disruption and with negligible crop losses. From a neutral stance, the ideal objective is to estimate the marginal benefits and costs of pesticide use at each use level. The marginal benefit at current use levels is likely to be low (pesticide use could be slightly reduced at little cost), while the marginal benefit at very low use levels is likely to be high (there would be high benefits in moving from zero chemical use to using some pesticides on high value crops highly susceptible to pest damage). Studies looking at total bans of all chemicals are far removed from the marginal valuation required and are completely unrealistic in terms of likely policy outcomes.

Some studies have made use of aggregate economy-wide models to estimate the effects of a reduction in agrochemical use on yields of various crops, and on farm incomes, employment, imports and exports, food prices, and so on. Many of these studies do not consider pesticides separately, and so are too aggregative to be of use in a chapter dealing only with pesticide use. However, it is possible in some cases to isolate the analysis of pesticides. In general, macro studies are unrealistic in that they ignore the possibility for the substitution of knowledge, labour, alternative products, or changed management practices, for reduced pesticide inputs. This assumption that farmers' activities will not change if pesticide use is severely restricted leads to overestimation of the reduction in agricultural output.

Knutson et al (1990) estimated the effects in the US of seven agrochemical reduction scenarios: no herbicides, no insecticides and fungicides, no inorganic nitrogen fertilizers, and various combinations. They assumed that reductions in chemical use occurred from 1991, and estimated the effects through to 1994, in comparison with a baseline scenario of business as usual. The crops analysed were maize, soybeans, wheat, cotton, rice, peanuts, sorghum and barley. The effects on poultry, pork, beef and the dairy sector were also assessed. These commodities account for over 75 per cent of the pesticides and over 70 per cent of the nitrogen fertilizer used in US agriculture. The estimates of effects on crop yields were made by a repeated questionnaire of 140 agricultural experts, rather than by actual experimentation, which must cast doubt on the results. Their estimates suggested, for the 'no chemicals' scenario, reductions in output ranging from 37 per cent for sorghum and soybeans, to 78 per cent for peanuts. This was predicted to lead to effects such as an increase in the rate of food price inflation from 4 per cent to around 13 per cent, and an increase in average household food expenditures by $428 per year for 1995 to 1998, leading to a total loss to consumers of $35 billion over the period. This is offset to some extent by increased farm incomes for crop producers, but incomes would fall for livestock producers due to increased grain prices. The finding that farm incomes increase when chemical use is reduced is a common one. This is because the increase in food prices when production falls is more than enough to offset the reduced production. The effects in the 'no pesticides' scenario were less extreme, with food price inflation rising to 8 per cent, and additional household food expenditure of $228 per year, adding up to a loss to consumers of $18 billion from 1995 to 1998.

Knutson et al were criticized by Ayer et al (1990) for overestimating the effect of reduced chemical use on yields and other economic impacts. Some of the main criticisms related to:

- failure to allow for the effect of new research and technological change in agricultural production if chemical use was restricted;

Table 3.4 *Effects on crop yield of pesticide and agrochemical reductions in Germany*

Crops	*Short-Term Effect (%)*		*Medium-Term Effect (%)*	
	'No Pesticides'	*'No Chemicals'*	*'No Pesticides'*	*'No Chemicals'*
Winter wheat	–28	–43	–35	–48
Barley	–29	–41	–36	–47
Summer wheat	–24	–40	–31	–43
Oat	–23	–35	–29	–36
Rye	–27	–42	–31	–44
Maize	–27	–39	–17	–27
Beans	–24	–29	–26	–26
Rape	–35	–45	–39	–52
Sugar beet	–35	–47	–37	–46
Potatoes	–41	–54	–48	–55
Maize	–25	–41	–31	–49
Pasture	–8	–30	–5	–24

Source: Schmitz et al (1993)

- insufficient allowance for improved conservation practices to conserve soil nutrients in response to higher crop prices and farm incomes; and
- the unrealistic assumption of total bans.

This last criticism is the most telling: the study derives its seemingly dire implications on the basis of the unrealistic scenario of 100 per cent agrochemical or pesticide reductions. This makes the results of the study of limited value with respect to analysing the costs of more reasonable changes in the use of pesticides and other chemicals.

Rendleman (1991) analyses the impact on the US economy of various reductions in all agrochemical use. The social cost of reductions, not including the environmental benefits, was found to be very low for reductions of up to 25 per cent, but then rose sharply, reaching around $3 billion for a 50 per cent reduction, $13 billion for a 75 per cent reduction, and over $25 billion for an 85 per cent reduction. Depending on the level of reduction, some or all of these costs would be offset by reductions in the costs of health care, health damage and environmental degradation. Because the Rendleman study focuses on equal reductions in all chemicals together, it cannot be used to infer the costs of pesticide reductions alone. However, the finding that the costs rise in a non-linear fashion with the amount of the reduction does confirm expectations that studies looking at 100 per cent or similar reductions will grossly overemphasize the costs of pesticide use reduction.

A different approach is employed by Chambers and Lichtenberg (1994). Existing studies of pesticide productivity (the MB function in Figure 3.5), which are often based on expert judgement, tend to estimate average productivity rather than marginal productivity, and are often wrong in retrospect; they opt for an econometric methodology applied to US agriculture as a whole. The basic approach is to estimate the 'production function' for pesticides (see Figure 3.1). Their analysis suggests, tentatively, that in the US crop damage may have been as high as 15 per cent of output in the 1950s, but that this damage declined to 11 per cent in the mid 1960s, 6 per cent in the mid 1970s and stabilized at about 3 per cent in the late 1979s, the change being due to the spread of pesticide use. Such reductions in crop damage are substantial.

Schmitz et al (1993) estimated the effects of four scenarios on agricultural output, crop and livestock markets, farm income and employment, and international trade for Germany. The scenarios were a continuation of the existing situation: 50 per cent reduction in nitrogen-based fertilizers, 100 per cent reduction in pesticides, and 100 per cent reduction in all agricultural chemicals. Table 3.4 shows yield reduction estimates for selected crops and scenarios. Like the study by Knutson, the Schmitz work suffers from unrealistic pesticide and chemical reduction scenarios. As a result, these estimates are once again of no use in suggesting the effects on crop production of any realistic reduction in pesticide use.

An interesting study of cancelling a pesticide is that for methyl bromide (MB), an ozone-depleting substance, in California. MB is widely used there and concern was expressed about the impacts on the economy of cancellation if an earlier (1996) rather than later (2001) date for cessation of use was adopted. One estimate put the losses in crop yields at $290 to $350 million per year, a loss arising in the main from the absence of immediately available alternatives (CDOFA, 1996). Moore and Villarejo (1996) challenge the study on the basis of poor and unsupported data, including substantial differences in figures on the use of MB in strawberry production, one of the main sectors alleged to suffer from MB withdrawal. The exchange illustrates the difficulty of verifying estimates for output loss from pesticide reduction.

Micro studies

Pretty (1993, 1995) has assembled convenient summaries of the evidence on returns from sustainable agriculture for Europe. Again the focus is on crop yields and profitability (gross margins), but the European context differs in that farm subsidies under the Common Agricultural Policy (CAP) are so large that output reductions are unlikely to have any impact on prices. However, if 'true' prices rise, the gap between intervention prices and actual prices could fall, thus reducing the size of government subsidy. In terms of crop yields, about 4 of the 16 case studies cited show increased yields from sustainable practice. Those that show declines are generally modest in their impact. In keeping with the US studies, gross margins generally increase rather than decrease as input costs decline.

From Theory to Practice: The External Costs of Pesticide Use

What kinds of externalities are involved in pesticide use? A typical list would include the following:

- health effects of pesticide poisonings;
- domestic animal poisonings;
- loss of natural predators;
- pesticide resistance;
- honeybee and pollination losses;
- crop losses;
- fishery losses;
- bird losses;
- groundwater contamination.

Table 3.5 *External cost estimates for US pesticide use*

External cost	*US $ million per annum*
Public health impacts	787
Domestic animal deaths and contamination	30
Loss of natural enemies	520
Pesticide resistance	1,400
Honeybee and pollination losses	320
Crop losses	742
Fishery losses	24
Bird losses	2,100
Groundwater contamination	1,800
Government regulations to prevent damage	200

Source: Pimentel et al (1992)

To this list should be added ozone-layer depletion from the use of methyl bromide. This section summarizes some of the studies that attempt to measure the external costs of pesticide use.

Comprehensive studies: US

A notable attempt to make a comprehensive estimate of the total external costs of pesticide use, for the US, is the study by Pimentel et al (1992) which builds on earlier studies (Pimentel et al, 1980; 1991a; 1991b). The results are shown in Table 3.5.

They estimate the external costs of pesticide use to be some $8 billion per annum for the US. This is in addition to the financial cost of some $4.1 billion. But the estimates for each category of damage are at best very rough. More problematic is that some of the methodologies are not economically sound. The cost item for groundwater pollution, for example, is based on the costs of clean up of groundwater, but it should be based on individuals' willingness to pay (WTP) for avoiding contamination. This willingness to pay could be less or more than the costs of clean up. If the WTP were more than the clean-up costs, it would be of net benefit to society to conduct the clean up; and if clean up actually occurred, then and only then would the costs of clean up be the appropriate measure of damages. However, where clean up is not in fact carried out, the correct measure of damages is the WTP to avoid them. In addition, any attempted clean up will be less than 100 per cent efficient, so there will be some residual damages (measured by the WTP to avoid them) which should be added to the cost of the operation. In any case, we need to estimate the full WTP to know whether or not a clean up programme is justified.

As a second example, the $2.1 billion estimate for the annual value of bird losses is also based on suspect calculations. An estimated 67 million bird deaths is multiplied by $30 per bird to give $2 billion. This $30 figure is an average of three other estimates. One was $800 per bird for replacement cost, which is not a valid estimate for the same reasons that the cost of clean up is not a valid estimate of groundwater contamination damage: replacement cost could only be used if replacement is in some sense absolutely necessary. A second estimate of $216 per bird was based on the expenditures of bird hunters, which relates to only a few game species, and in any case may reflect neither the hunters' nor the public's marginal valuation of bird losses. The third estimate of

$0.40 is based on the costs of birdwatching, which is equally invalid. None of the estimates reflects individuals' valuations of bird deaths, and the $30 per bird figure must be seen as ad hoc (Bowles and Webster, 1995). The remaining $100 million of the $2.1 billion figure is based on the annual expenditures of the US Fish and Wildlife Service on its Endangered Species Program, and while some of the species in question suffered heavy losses from early pesticides, in particular those causing shell thinning in birds of prey, there is little or no direct connection between the bird damages of current pesticide use and the expenditures of the programme. If replacement cost was a valid measure of WTP, then, by definition, damages would always equal costs: in other words benefits would always equal costs.

Thirdly, the estimates for resistance are problematic because many of the effects are borne by farmers themselves. Unless they are wholly ignorant of the impacts of using pesticides, farmers will effectively have internalized these costs. They are not externalities. In practice, it seems likely that there are large areas of unawareness so that farmers do not appreciate the full resistance effects of pesticides. Thus three of the largest category estimates in Pimentel's figures are flawed in terms of the underlying theory of externalities. There are similar concerns for many of the other categories: some figures are likely to be overestimates, while others may be underestimates. A further concern is that some figures relate to the misuse of pesticides in quantities beyond the recommended levels, rather than to pesticide use as such. The social costs thus appear to be made up partly of 'true' costs and partly of the costs of mismanagement. From a policy standpoint, the latter may be better addressed through information campaigns, rather than through measures designed to restrict pesticide use in general.

Although the estimates made by Pimentel et al are rather rough and unreliable, for some categories they are the only estimates in the literature. Pimentel himself views the figures as a starting point on which others may improve. There is indeed much scope for improvement, in particular for health damages, where a wider range of estimates can be found and a fuller range of countries considered. The remaining problem is that the Pimentel estimates are of limited use in deriving any policy implications because of their focus on total costs, and because no distinctions are drawn between the different chemicals used.

A second attempt at an overall analysis of US pesticide externalities is that of Steiner et al (1995). This study measures some of the categories of externality identified in the Pimentel studies but omits domestic animal deaths (minor in the Pimentel study), pesticide resistance (a major item in the Pimentel study – see Table 3.5) and crop losses (also a major item). The rationale for excluding most pest resistance costs is that they are borne by farmers in the form of reduced yields and are therefore internalized (see the discussion on Pimentel above). Estimates for pollination losses, regulatory costs and even overall biodiversity loss are similar to the Pimentel study (although the biodiversity damage figure ranges from $0.3 billion to $20 billion). There are substantial differences for accident costs but the basis for their estimation is difficult to identify in the paper. Chronic health effect costs are estimated to be the same as acute effect costs but no rationale is provided for this assumption. Health costs associated with contaminated water are estimated as the cost of the regulations, a fallacious procedure for estimating damage which should be based on willingness to pay to avoid contamination (see below). The end result is a range of $1.3 billion to $3.6 billion for externalities, two to six times less than the Pimentel study.

As a thought experiment, we can imagine that the Pimentel and Steiner analyses are credible. Let the benefits of using pesticides, measured by crop protection, be four times the expenditure on pesticides (see earlier discussion). Then, expenditures in the

US in the two studies are $4 billion and benefits $16 billion. Externalities are $8.1 billion in the Pimentel study and, taking the maximum, $3.6 billion in the Steiner study. A benefit-cost ratio is measured by:

$$B/[C + E]$$

where B is benefits, C is expenditure on pesticides and E is the externality level. The benefit-cost ratio would be 16/[4 + 8.1] = 1.32 for the Pimentel case and 16/[4 + 3.6] = 2.10 for the Steiner case. Since the ratios are greater than unity (benefits exceed costs), it is tempting to conclude that the prevailing use of pesticides is justified. But no such conclusion follows because the optimal use of pesticides is where the benefit-cost ratio is maximized, and this could be at a point where more or less pesticides are used. Moreover, high benefit-cost ratios tell us little because the policy issue is whether there is some alternative 'pesticide technology' package, say integrated pest management or low-input agriculture, which will secure an even higher benefit-cost ratio. Thus, interesting though such 'global' damage studies are, few policy conclusions follow from them.

It is also notable that neither study includes any beneficial externalities from pesticide use. The general rule is that the benefits will be internalized in the price of pesticides and hence there are no external benefits. But one candidate might be the effects of reducing pesticide use on overall land use. Suppose, for example, that the effect of reducing pesticide use is to lower crop yields. One effect might be to bring into production currently unutilized land in order to expand production. Since unutilized land could generally be assumed to have a higher environmental profile than agricultural land, the effect could be environmentally damaging. The question, however, is whether this effect is likely, and even if it is, whether it constitutes a positive externality. In developed economies land uses tend to be in an equilibrium such that land is not converted further for agricultural use. In less developed economies this does not hold and there are pressures to bring 'wildlands' into production. If such land expansion is feasible, the issue is whether an externality is involved. If land uses were in equilibrium then, at the margin, agricultural land would have the same value as non-agricultural and potentially convertible land. If a pesticide reduction policy lowers yields it will also lower agricultural land values, making it less, not more, likely that non-agricultural land will be taken into agricultural use. While the issue is not straightforward, we conclude that there is unlikely to be a positive externality from pesticide usage.

Health damages: Direct and indirect exposure

There are two principal strands of health damage to be considered. One is acute pesticide poisoning, which occurs to individuals exposed to high levels of pesticides for a short time. This category covers isolated incidents such as when treated seed grain is accidentally used for consumption, or when regulations regarding harvest intervals are not observed. It also covers cases in which pesticide applicators, or members of the public, are exposed to high levels of pesticides during or following the application of chemicals. These effects are easier to link back to pesticide exposure, and data are available on acute pesticide-related illnesses and deaths for many countries. Nevertheless, it is likely that acute effects are under-reported in many cases, partly because people may ignore symptoms unless they are particularly severe, and partly because medical personnel may often wrongly diagnose pesticide poisonings as being some other, more common, illness with similar symptoms.

One complex issue relating to health damages is the extent to which occupational exposure is a true externality. The economic model previously outlined assumed not only that farmers alone make pesticide use decisions, but that farmers ignore the external costs, including health and safety aspects. Beach and Carlson (1993) investigate herbicide choice and suggest some partial evidence that farmers actually bid up the price of safer pesticides: some of the health externality is internalized in farmers' decisions. If so, then not all occupational health exposure costs can be regarded as externalities (see also Angehrn, 1996). As noted earlier, evidence from developing countries suggests an awareness of risk but limited understanding of the pathways by which pesticide contamination can occur, and a limited response in terms of protective measures undertaken.

Such evidence as is available suggests that there may be as many as 25 million cases of acute pesticide poisoning among developing world agricultural workers alone each year (Jeyaratnam, 1990). The Helsinki Institute of Occupational Health reports that 11 million cases of acute pesticide-related poisoning occur each year in Africa, a continent where 80 per cent of the population is involved in agricultural related activities; but most of these victims do not seek medical help and therefore are not officially documented in government reports. About 50 per cent of all pesticide poisonings, and 80 per cent of pesticide-related deaths, occurred in developing countries, up to the mid 1980s (Pimbert, 1991). This situation seems likely to continue, with pesticide use falling in the developed world, but still rising rapidly in most of the developing world (see Table 3.6). The WHO (1990) estimates 40,000 *unintentional* deaths from pesticide poisoning each year.

In both the developed and the developing world, many acute poisoning incidents are the result of suicide or parasuicide. Many other cases arise through other forms of inappropriate use of the chemical, or through failure to observe basic safety requirements. As noted above, it would be improper to count these costs as part of the external costs of pesticide use, because the costs could be reduced by better information or tighter controls, without the need to reduce pesticide use as such. However, such reductions in the costs may not always be possible, especially in developing countries, where the resources for extensive education or control campaigns may not be available. Thus, reducing the extent of easy access to highly toxic chemicals such as paraquat could result in the saving of many lives. Other reasons for unsafe pesticide use in developing countries include widespread illiteracy, a lack of protective clothing and equipment, a lack of adequate medical care and other professional services, and a lack of training and safety procedures. The costs arising from the misuse of pesticides will remain until these root problems are addressed.

Nonetheless, while these costs demand consideration and remedial action, they should not be considered as part of the external costs of pesticide use. The costs of misuse are real costs suffered by society, but by keeping them separate from the analysis of the private and external costs of pesticide use, the distinction between the different policy implications of the different costs is maintained. The other form of health damage, chronic health impairment, arises through prolonged exposure to low levels of pesticides. The most significant effects can be expected for those working in the pesticide and agricultural industries. But long-term exposure also occurs for the general population, through contamination of groundwater used for drinking, cooking and washing, and through residues of pesticides found on food.

The California Department of Pesticide Regulation (1993) classifies health impacts related to pesticides as 'definite', 'probable' and 'possible'. Taking the definite and probable categories only, there were 1225 cases of occupational related illness and injury in 1990, and 36 non-occupational cases. The dominant categories of illness were systemic eye and skin disorders. In addition, four 'definite' deaths linked to pesticides were recorded, one of which was a suicide. Reus et al (1994) find that little is known about the chronic effects of pesticides in the European Community, but they cite a French study of 4000 farmers in 1985 in which 18 per cent of pesticide users cited headaches, eye problems and digestive troubles.

In a study of farmer groups in the Philippines, Antle and Pingali (1994), Pingali et al (1994) and Pingali et al (1995) found eye, pulmonary, dermal, neurological and kidney problems to be statistically associated with long-term exposure to pesticides. The authors aggregate the health impairment by estimating the medical costs of treatment and the forgone output during the recuperation time – a 'human capital' approach to valuation. Average health costs for farmers exposed to pesticides were found to be 60 per cent higher than health costs of farmers not exposed. This approach has the advantage of producing economic costings, but it should be pointed out that the link between human capital measures and true WTP is not systematic and the human capital approach is a largely discredited approach in economic valuation. Nonetheless, the study is interesting because it enables the productivity losses from pesticide exposure to be compared directly with the productivity gains in terms of crop protection. The authors find that the health benefits offset the crop protection losses if insecticides are reduced, but that the health damage from herbicides was not significant, so that herbicide reduction has a net loss of overall productivity.

Most health studies focus on cancers and poisonings, but Repetto and Baliga (1996) suggest that the focus of concern should be widened to include suppression of immune systems. While some immunosuppression may occur in healthy humans without significant effect, this is unlikely to be true in populations already at risk from malnutrition, waterborne diseases and so on. If this is true, then the health impacts of pesticides are far more widespread than previously thought.

Health damages via food consumption

More problematic still is the link between pesticide residues in food and morbidity and premature mortality. According to a study by Chemishanksa (1982), 90 per cent of human pesticide intake is through food. The extent of these damages is difficult to estimate, because it is hard to uncover the long-term causal linkage between small levels of pesticide exposure and effects such as cancer. In addition, the rapid development of new chemicals with different properties means that even very good data on the long-term effects of exposure to existing pesticides may be a poor predictor of the effects of the next generation of agrochemicals. In assessing the risk associated with regular exposure to low levels of pesticides, we have to rely on studies of animals, mostly rats and mice, exposed to higher levels for shorter periods. This can give only a very rough idea of the chronic risk in humans.

In 1992, about 35 per cent of food samples in the US contained detectable residues, but only 1 per cent contained residues at concentrations greater than US-EPA tolerance levels. The Natural Resources Defense Council (1992) assessed the exposure of American children under five years of age by estimating food consumption in this age group together with the likely content of 23 pesticides in that food consumption. They concluded that between 5500 and 6200 children might develop cancer solely as a result of exposure to pesticides through this food link of fruit and vegetables. Taking a US value of statistical life (see below) of $2 million, this would suggest an economic value of some $12 billion. This is not an annual cost because the estimate relates to the

Table 3.6 *Annual health impact of pesticides in developing countries*

Population Groups at Risk	*Character of Exposure*	*Estimated Overall Annual Impact*
Pesticide formulators, mixers, applicators, and pickers; suicides; mass poisoning incidents	Single- and short-term, very high-level exposure	3,000,000 exposed; 220,000 deaths
All population groups	Long-term, low-level exposure	37,000 suffer from chronic effects of long-term exposure (such as cancer)
Pesticide manufacturers, formulators, mixers, applicators, and pickers	Long-term, high-level exposure	735,000 suffer from specific chronic effects of long-term exposure

Source: WHO (1992)

entire 'stock' of American children at one point in time. Nonetheless, it is a substantial figure. In the European Community little appears to be known about the risks from food contamination. Maximum residue levels (MRLs) were found to be exceeded in 2 per cent of food samples in the UK, 3 to 7 per cent in Germany, and 14 per cent in Italy (with estimates of 48 per cent in oranges) (Reus et al, 1994), but no estimates of excess morbidity, if any, from these concentrations appear to be available.

Public perception of pesticide risk suggests that the public, in the US at least, has a high sensitivity to this source of health risk. Horowitz (1994) reports a random survey of US households which shows a distinct preference for pesticide regulation over other sources of risk even where the costs of control are the same and the same number of lives is saved. This observation could be due to the greater scientific uncertainty about pesticides, and to the idea that pesticide exposures are less observable, and therefore less under the control of the individual.

Willingness to pay for pesticide-free food

Given that the external costs of pesticide use are extremely difficult to identify and measure, an alternative approach is to look at the premia over the prevailing market price that individuals would be willing to pay for pesticide-free, or pesticide-reduced, produce. Such an approach captures at least part of the overall level of externality, namely that part relating to public health risks as perceived by the consumer. It may or may not capture the risks to the environment generally. It will do so if consumers are motivated to pay a premium because of general concern about the role of pesticides in the environment, something that can be tested in well-designed questionnaire approaches (this has been tried in Chapter 4). A similar possibility exists with respect to worker-health risks: consumers may be willing to pay a premium to ensure that such risks are minimized. While it is popularly assumed that 'economic man' is selfish and would not have such altruistic motivations, there is, in fact, nothing in economics that requires willingness to pay to be selfishly motivated.

Buzby and Skees (1994) found that respondents in a survey had a mean willingness to pay for 'safer grapefruit' (50 per cent or 99 per cent risk reduction from pesticide residues) of \$0.15 to \$0.69, over a basic price of \$0.50. Higher willingness to pay was expressed for higher risk reductions, and by females and those voicing strong concern about pesticide residues. Weaver (1992) surveys various sources of evidence on the link between pesticide use and consumer demand for produce quality. Evidence from *The Packer* shows that there is a relatively stable and significant consumer concern about pesticide residues, with around 75 per cent reporting concern. Around 20 per cent have changed consumption decisions due to these concerns, buying organic food or avoiding treated fruit and vegetables. Weaver quotes from work by Ott (1990): 65.5 per cent of respondents in a survey were willing to pay 5 per cent more for 'certified pesticide residue free' produce. He quotes from a Gallup poll in 1989 the findings that 50 per cent of respondents would be willing to pay more for organic food, with 41 per cent being willing to pay 10 per cent more, 16 per cent being willing to pay 20 per cent more, and less than 10 per cent being willing to pay more than 20 per cent. Weaver et al (1992) found that consumers were generally willing to pay more for pesticide-free produce (tomatoes) although concern was expressed at cosmetic defects in pesticide-free produce. Willingness to pay results were:

Willing to Pay:	*Percentage of Sample*
No more	19%
up to 5% more	25%
5–10% more	30%
10–15% more	26%

Baker and Crosbie (1993) look a little further into this issue, using techniques which enabled them to identify three stable 'clusters' of consumers with different attitudes towards apple purchase decisions. The first, comprising 16 per cent of consumers, was the most concerned with pesticides, in particular because of suspected carcinogenic effects. By comparison, they were not worried about appearance or price. The second cluster, with 55 per cent of consumers, valued cosmetic damage most and were relatively insensitive to price. The remaining 29 per cent were sensitive to both quality and price. The results can be used, for each cluster, to predict the value of changes in pesticide-residue levels. For each cluster, a function was estimated relating total utility, or benefit, depending on various characteristics of the apples: price, pesticide residues, cosmetic appearance, and so on. By calculating the change in this benefit per one cent change in the price paid, and the change in the benefit when pesticide residues change, a comparison can be drawn between pesticide changes and price changes. One important conclusion from the Baker and Crosbie work is that the preferences of each one of the three clusters were poorly approximated by an average function estimated for all the consumers together. The implication is that it is important to differentiate among different clusters of consumers when analysing preferences, which significantly increases the complexity of work in this area.

Eom (1994) conducted a sample survey to test the link between consumer perceptions of risk and willingness to pay premia. He found significant willingness to pay for very small risk reductions but willingness to pay was also invariant with respect to changes in the level of risk. He suggests this may be because consumers were valuing food safety as such rather than specific risk levels. Probably the most extensive studies are those of van Ravenswaay and coworkers who have investigated attitudes to pesticides in food and consumers' willingness to pay to avoid residues (see van Ravenswaay and Hoehn, 1991a, 1991b, 1991c, 1991d; van Ravenswaay et al, 1992; van Ravenswaay and Wohl, 1995; and Wohl et al, 1995). Van Ravenswaay and Hoehn (1991d) tested an approach in which apples were labelled with different levels of pesticide residue. Using contingent valuation, they sought consumers' willingness to pay to reduce residue levels. They found that consumers could discriminate between risk levels and were willing to pay positive amounts for lower residue content. However, risk perceptions among consumers appeared not to account for the full effect of the labels on consumer demand. Consumers were therefore 'buying' something else besides risk reduction when they purchased the lower risk (but higher price) apples. Van Ravenswaay and Hoehn suggested that what was being purchased was reduced *ambiguity* of risk. Risk ambiguity refers to a situation in which the probability distribution is itself not known with certainty. Rather than the risk being, say, 10^{-6}, it is a range from 10^{-9} to 10^{-3} with a mean of 10^{-6}. The literature suggests that people are willing to pay larger positive amounts to reduce this second order probability than to reduce the probability of risk known with certainty. Put another way, people prefer a context in which the risk is known to one in which the risk is itself uncertain.

In a follow up survey, van Ravenswaay and Wohl (1995) modified the original questionnaire in several ways, offering consumers the choice between labelled and unlabelled apples. Labels consisted of 'produced without pesticides' and 'no pesticide residues above federal limits'. The questionnaire was applied by telephone to a sample of Michigan residents in 1992. The perceived risk was found to have a mean of between 10^{-4} and 10^{-3}, but the dispersion was wide: consumers simply have very different perceptions of risk. Risk ambiguity was also measured by seeing how certain people reacted to risk (probability) estimates: only 23 per cent were 'sure' while 45 per cent were 'somewhat sure'. Over 50 per cent of respondents thought the associated health risks were risks of cancer. Preference was shown for reducing risks to 'produced without pesticides' levels but not markedly so above the 'meets federal standards' level: there was a perceived feeling of assurance about federal regulations.

By integrating risk ambiguity into a willingness to pay study using contingent valuation, Wohl et al (1995) estimate that Michigan residents are willing to pay 32 per cent more per pound of apples that meet certification standards, or, multiplied by the average consumption per year, around $7 per household per year. Buzby et al (1995) find a similar incremental willingness to pay for pesticide-free grapefruits using the 'payment card' approach to contingent valuation, and a 134 per cent increase when the dichotomous choice format is used. The policy implications of the van Ravenswaay work are important: reducing risk ambiguity can have significant social gains and this in turn suggests that labels guaranteeing that regulations are met, and assurance that enforcement will occur if they are not, are important. More generally, the studies indicate that risk studies have in the past paid too little attention to risk ambiguity.

The van Ravenswaay studies have also considered the extent to which consumers are willing to accept scarred and blemished produce as the price of pesticide-free produce. They conclude that, in general, consumers are not willing to accept produce that offends some reference perception of what the produce should look like, just because it is pesticide free. This finding is consistent with other studies.

Caswell (1995) brings together a number of studies on willingness to pay for food safety ranging from fatty acids in fats and oils, contamination of shellfish, pesticides in fruit, to salmonella in meat and chemical residues in sportsfish. The overwhelming conclusion of these essays is that there is a positive willingness to pay to reduce food risks. All these studies have policy relevance, but it is limited to the willingness to pay for consumer safety only. There is scope for wide-ranging work looking at individuals' direct valuations of the more general effects of pesticides, moving away from the food focus. The studies also tend to rely on fairly small samples. Perhaps the most important limitation is that the studies tend to look at 'pesticides' in general, rather than at individual chemicals. Thus the results are of limited help in determining choices between pesticides.

Valuing health damage

Despite the uncertainties involved, and the wide range of potential estimates, the available figures help to provide some indication of the overall scale of the pesticide-damage cost problem. Even ignoring suicides and non-fatal illnesses, the estimated 40,000 unintentional deaths per annum is a substantial figure. What value should be attached to such premature mortality ? The relevant figure is the 'value of a statistical life' – a VOSL.

The way a VOSL is obtained is by aggregating up from a value (willingness to pay) of risk reduction. Imagine the probability of dying next year is 0.004 for each person and suppose we have 1000 persons in the population. Assume there is some risk-reduction policy that reduces the risk to 0.003, a change of 0.001. Each person is asked to express their WTP for this change in risk. Suppose the answer is £1000. The risk-reduction policy is a public good: it affects everyone equally. Thus 1000 people say they are each willing to pay £1000 for the policy: in other words their aggregate willingness to pay is £1 million. The change in risk will result in one statistical person being saved each year (1000 x 0.001). Thus the value of a statistical life is £1 million in this example. It is important to understand that no one is being asked their WTP to avoid dying at a specified time: they are being asked to express a WTP for a change in risk. These values can be expected to vary across different individuals. The two main reasons for this will be that:

- People have differing attitudes to risk: some may even be risk lovers positively enjoying risky contexts. Most people are risk avoiders; they will tend to reveal a positive willingness to pay for risk reduction. But there is no particular reason why their valuations of risk should be the same.
- Incomes vary, and hence willingness to pay is likely to vary in such a way that those with higher incomes have higher WTPs; but this is not a necessary result since attitudes to risk may vary in such a way as to offset an income effect.

A VOSL can also be measured by a 'willingness to accept' (WTA) compensation for increased risk. It is well known that many people do make this trade-off between risk and money, for example by accepting premia on wages to tolerate risk. It is tempting to think that the WTA approach will produce very much higher values for a VOSL than the WTP approach, simply because WTA is not constrained by income. WTP and WTA can, indeed, be different and WTA for environmental losses may exceed WTP for environmental gains by factors of two to five. Various explanations exist for this disparity, including the fact that individuals may feel they are losing a property right if the issue is one of loss of an asset (WTA) rather than an increment to an existing property right (WTP). Another explanation, which is wholly consistent with economic theory, suggests that WTA > WTP arises mainly in contexts where there is no ready substitute for the environmental good in question.

The evidence on WTP and WTA for risk is far from conclusive on this issue. Pearce, Bann and Georgiou (1992) review the various estimates of VOSL and find the mean estimates across studies given in Table 3.7:

Table 3.7 *Value of a statistical life estimates*

	m (1991)	
	USA	*UK*
WTA (wage risk)	2.5–3.9	2.0–2.5
WTP		
(contingent valuation, contingent ranking)	1.0–1.8	2.9–4.5
(market)	0.7–0.8	0.5–2.4
WTA/WTP		
(wage risk/contingent valuation)	1.4–3.9	0.4–0.9
WTA/WTP		
(wage risk/market)	3.1–5.6	0.8–5.0

Source: Pearce, Bann and Georgiou (1992)

This suggests that the WTA > WTP inequality holds for the US but not for the UK.

Evidence on VOSLs in the developing world is scarce. The few studies that are available suggest very low VOSLs relative to the VOSLs for North America and Europe, perhaps of the order of $10,000 to $50,000. This is partly as expected since income differences are substantial and WTP depends on income. On grounds of fairness it might be preferable to adopt the same VOSL regardless of where the risks occur. While there is some ethical force in this view, it is not consistent with the way life-saving expenditures are actually made. It is unlikely, for example, that the expenditure per life saved is the same in developed countries' foreign aid budgets as it is for their own health services.

A further problem is that much of the risk from pesticide exposure in the developing world is concentrated in a relatively small group of agricultural workers, regularly exposed to pesticides, each of whom faces a relatively large risk. It is likely that willingness to pay to avoid a given risk is not a linear function of that risk, but rather that, over the range in question, the average willingness to pay per unit risk falls as the total risk rises. Adopting a notional $100,000 per life saved, the annual premature mortality cost for the world as a whole would be of the order of $4 billion.

Surface water and groundwater contamination

Various studies exist on the willingness to pay to avoid groundwater contamination. Pesticides are one of the sources of such contamination and a US Office of Technology Assessment report concludes that pesticide residues are in groundwater in virtually all states, and that drinking water standards are violated in states where there is heavy use of pesticides (US OTA, 1995). Some surface waters are also contaminated, particularly by herbicides in contexts where reservoirs have low recharge rates, limiting the flushing out of pesticide residues. In such contexts damage is effectively irreversible since the pesticide becomes an accumulative pollutant.

Edwards (1988) conducted a WTP analysis for groundwater protection in the US, where the risks to groundwater were from nitrates rather than pesticides. The 'present value' (the discounted value of future benefits) was from $5 million to $25 million per 1000 households, depending on the probability that supply would be achieved. The upper figure relates to a certainty that water would be supplied from the source. The per household WTP is thus $5000–$25,000 in present value terms, or, say, $400 to

$2000 per annum at a discount rate of 5 per cent and an assumed time horizon of 20 years. This suggests that groundwater quality is highly valued. Later studies suggest much lower figures. Smith (1992) uses willingness to pay estimates for clean groundwater to derive a figure of $800 million for groundwater damages from pesticides and nitrates in the US (in 1984 prices). This figure is made up of $33 per household per annum multiplied by the number of potentially affected households (32.6 million) to obtain an aggregate value a little over $1 billion, which he then adjusts downwards to $800 million to account for household response rates to the original survey. Crosson (1995) suggests that the figure is an exaggeration because the number of households actually exposed to such risks is very much smaller.

A survey of groundwater contamination in the European Community found that specific pesticide concentrations exceeded the EC Drinking Water Directive guidelines of 0.1 micrograms per litre in a number of countries. Simazine and atrazine, for example, were found to exceed the guideline in six countries (Reus et al, 1994). Data on concentrations in European surface water are scarce. The same report suggests that atrazine, simazine, molinate, metolachlor and alachlor – all herbicides – are the most commonly encountered, often exceeding the maximum tolerable concentration (MTC). US surveys have found that less than 10 per cent of surface water samples contain detectable levels of 11 organochlorine insecticides, seven organophosphate insecticides and four herbicides.

Soils

Pesticides in soils tend to kill soil microbes and alter microbial food webs in the soil. This may actually make crops more susceptible to disease and reduce crop growth – for instance, by reducing nitrogen fixing capacity and thus higher expenditures on fertilizer. No reliable estimates appear to exist of the monetary cost of soil damage from pesticides.

Biodiversity

Apart from health damages, probably the largest concern about pesticides relates to their impact on biodiversity. Yet this is the subject about which least appears to be known, at least in quantitative terms. Two steps are required for any rigorous analysis. Firstly, a 'dose response function' linking pesticide applications to diversity and single species abundance is required. Secondly, some idea of the economic value of diversity is needed. In practice, neither step has been addressed adequately in the literature. An example of a micro dose response relationship, though not of a quantitative function, is provided by Rands and Sotherton (1986) who investigated butterfly concentrations on arable land, part of which had been sprayed and part of which had not. Over 800 butterflies were recorded on the unsprayed plot and less than 300 on the sprayed plot. Reus et al (1994) note that biodiversity loss is most likely in areas where spraying occurs directly, but that indirect effects from volatilized pesticides transported over wider areas may be extensive, though poorly monitored.

Pollinator damage appears to be a little better understood. The economic value of honeybees to US crops has been estimated at between $4.5 and $18.9 billion per annum. Pesticides are a major cause of bee mortality. Pimentel et al (1992) suggest that lost honey production and the cost of hiring hives to compensate for natural population losses arising from pesticides could cost the US some $120 million per annum. They also offer a rough estimate of a further $200 million loss from reduced pollination. Part of the problem lies in the specificity of effects. For example, insecticides may deplete soil fauna and thereby harm processes of decomposition and nitrogen fixation. But the relationship between dose and response in such a context is highly complex and dependent on many different conditions. Generalized dose response relationships are unlikely: a quantitative relationship for one area almost certainly will not apply to other areas. The analysis of biodiversity externalities, therefore, remains unsatisfactory from the standpoint of economic analysis.

Resistance

It is well known that continued use of pesticides builds up resistance in some target pest populations. The number of resistant species has also increased dramatically this century (Georghiou, 1986). One response to this is to increase pesticide usage, thus giving rise to the pesticide treadmill where the reduced effect of pesticides results in more pesticide use. As an example, Imperial Valley, California, cotton growers used 74 per cent more organophosphates in 1981 than in 1979 to kill the pink bollworm, with an estimated return of just $1.14 for each $1 spent on pesticides (note that this 14 per cent rate of return can be compared to the 4:1 ratio referred to above under the benefits of pesticide use) (Walker et al, 1995). A further response is to switch out of the reduced-productivity pesticides into newer varieties. These may lower some of the externality risks since new products tend to have tighter environmental and health regulations attached to them. There are, however, exceptions, as with the reduced effectiveness of the rodenticide Warfarin which was replaced with newer anticoagulants of greater risk to humans. Pesticide use may also result in secondary pests: pests that previously did little harm but which multiply as a result of their predators being targeted by pesticides. Examples include aphids, scales and whiteflies.

Pesticide use can involve two kinds of externalities between farmers. Destruction of pests by one farmer contributes to reduced damage on adjacent farms – an example of a *positive externality* (a benefit). But continued use of pesticides also produces pesticide resistance so that adjacent farms suffer the consequence of neighbours' use of pesticides. Kishor (1992, 1993) cites the example of cotton crop devastation in two districts of Andhra Pradesh in India in 1987–88 due to excessive pesticide use. Moreover, resistant strains may migrate to other crops, further intensifying the negative externality. Kishor's study of Andhra Pradesh provides one of the few reasonably well-researched estimates of damage from pesticide use. Externalities are classified as continuing annual losses and probabilistic catastrophic losses, and are further differentiated as externalities on crops other than cotton in the relevant district (Guntur), and losses outside the district. Kishor suggests the following results:

- average external cost of cotton cultivation: 26 to 104 per cent of average private cost;
- probability of catastrophic loss is one in seven years, with 50 per cent expected crop damage to pulses and an annual cost of $8 million, or a further increase in private cost of 10 per cent.

Kishor's study also illuminates the benefits of IPM and some of the costs of overusing pesticides. He estimates one benefit of IPM is a reduction in pesticide expenditure by over 20 per cent. Added to this, IPM increases yields, so that the total benefit of reduced expenditure and increased yields comes to a reduction in average costs of cotton production of some 28 per cent. Just the 20 per cent expenditure reduction would yield some $5 million savings in the Guntur district. Adding the increased yields because of IPM, and deducting the extra labour costs of IPM, would have the effect of raising net revenues by an amount equal to just under 5 per cent of the entire district's income, a remarkable rate of return. In terms of the externalities, IPM is estimated to reduce the probability of catastrophic damage to one in 28, thus reducing these costs further. Kishor's work is important as one of the few attempted quantifications of some of the externalities associated with pesticide use.

Non-Use Values

Non-use values (sometimes called 'passive use' values) are expressions of willingness to pay to conserve or avoid something even though the 'valuer' has no direct experience of the object in question and does not intend to have any. There is a controversy in environmental economics as to whether non-use values legitimately belong to cost-benefit appraisals. One argument is that they may reflect some form of commitment or altruism and that this does not fit easily with the assumption of self-interested behaviour. However, there is no restriction that suggests willingness to pay is motivated by self-interest alone, so this criticism appears to be misdirected. No study of non-use values for pesticides appears to exist. Bowles and Webster (1995) cite the study by Higley and Wintersteen (1992), but there is no evidence that farmers' willingness to pay for reducing the environmental consequences of pesticide use is unrelated to their own enjoyment of the benefits.

The Cost-Benefit Equation

What can be concluded about the extent of external costs from pesticide use? An exercise such as that by Pimentel and colleagues (1992) deserves praise for bringing the issues into focus through the medium of 'monetization'. Nonetheless, the study itself is not always credible because of the absence of a proper economic methodology and the major questionable assumptions on which a number of the estimates rest. Closer inspection of the literature on the externalities reveals extensive uncertainty about the effects of pesticides on, for example, biodiversity, one of the probable major impacts. At best, then, we suggest a 'thought experiment'.

The annual value of pesticide sales worldwide is some $31 billion. The limited empirical evidence suggests that the productivity of pesticides may range from one to five: the total benefits of pesticides in terms of avoided damage are some $31 billion to $155 billion. If benefits and costs are equal at $31 billion and if external costs are positive, as they surely are, then a case exists for a major reduction in pesticide use and a switch to alternative means of control based on sustainable agriculture practices. Figure 3.6 illustrates this point on the basis of a reasonable expectation of the shape of *total* benefit and cost functions. If the productivity of pesticides is unity, then the position is similar to 'A', which clearly involves excessive pesticide use.

If productivity levels are higher, at, say, four or five as suggested in some of the literature, then the position is closer to a point such as B where private net benefits are maximized. We cannot be sure that the economic system is at B because the ratio of 4:1 or 5:1 will also hold at other points on the diagram to the left of A. But if pesticide use is 'rational' and farmers seek to maximize profits, B is a reasonable expectation. The ratio BE/DE would then be 4:1 or 5:1. But this is not a socially desirable level of pesticide use because external costs (distance CD) remain. Socially desirable use is at F,

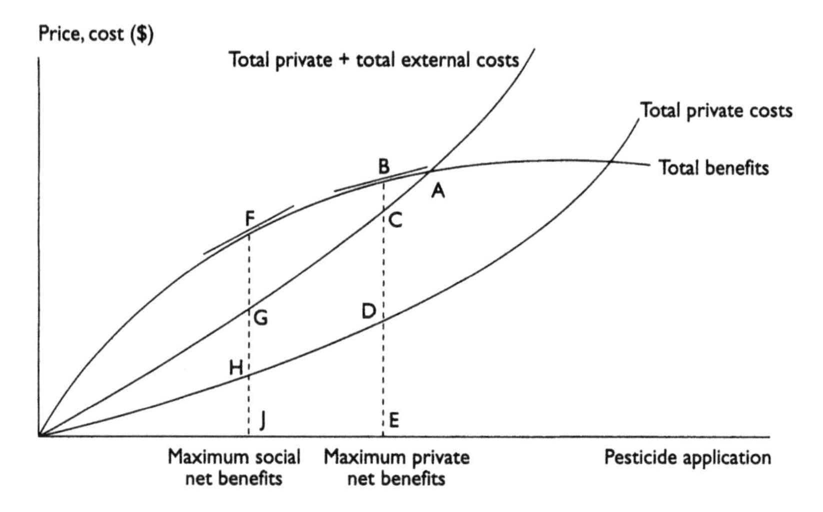

Figure 3.6 *Socially optimal pesticide use*

where the ratio of private benefits and costs could still be 4:1 or 5:1 (FJ/HJ), but where the external costs (GH) are accounted for, though not eliminated as previously explained.

There are several reasons for supposing that the prevailing situation is to the right of a point such E. Firstly, as noted, the hypothesis of profit maximization by farmers puts us somewhere in the region of point B rather than E. Secondly, we know that while some externalities may in fact be internalized (some of the health damage to farmers), much of it is not. Hence the system must be operating to the right of E. Thirdly, we know that, while many subsidies have been reduced in recent years, many remain. This has the effect of lowering the total cost curve in Figure 3.6 and hence making profit maximization achieve a position even further to the right of E.

However, the problem is just how large the reduction in pesticide use should be. If Pimentel et al (1992) were right in their estimates, then, for the US, total social costs would be of the order of $12 billion against total benefits of $16 billion. Benefits would still exceed costs, but if the $8 billion is not internalized, substantial reductions in pesticide use would be called for. In terms of Figure 3.6, BE = 16, DE = 4, and CD = 8, assuming the actual situation corresponds to the profit maximizing situation. Little can be said about the scale of the reduction in use that is required since we do not know the benefit and cost *functions*, but it is reasonable to suppose it would be substantial. But the Pimentel estimates are not reliable. In their place we have a very uncertain knowledge of the biodiversity effects and a health cost that certainly runs to billions of dollars. Induced cancers in the US, for example, might run to several billion dollars per year just for childhood exposure.

No quantitative estimate of the required reduction in pesticides appears to be possible, but the direction of control is clear. That the scale of the reduction required is large is also evident given the existence of alternatives packages such as IPM which, the evidence suggests, may not reduce yields and may even reduce costs, regardless of any environmental benefits. While there appears to be some consumer resistance to the side-effects of reduced pesticide use in terms of the cosmetic appearance of food, there also appears to be a clear preference among consumers for reduced pesticide use.

The extent to which these findings can be extrapolated from their origin, usually the US, to other countries is open to question. But as developing nations grow and become richer, their preferences may also converge on demanding better product quality.

In other respects, the decision context is far more straightforward. The issue is generally not one of using an amount X of pesticides or using, say, 0.5X and forgoing the profits from the pesticides not used. Rather it is one of using pesticides as a standalone technology or using different kinds of pesticides in a package of integrated pest management (IPM) measures. IPM packages would include biological pest control, biologically based pesticides, pest resistant crops, and traps for luring pests. In practice, many IPM systems use conventional pesticides, so that IPM does not *necessarily* reduce pesticide use. Nonetheless, reviews by Carlson and Wetzstein (1993) and Norton and Mullen (1994) suggest that IPM has lower overall costs without a comparable loss rate – in other words, increased profitability, plus reduced pesticide usage. IPM has extensive use in some countries, including the US. If it is so cost-effective, why is it not used more widely? One suggestion is that, while overall costs are lower, management costs and learning costs are higher, and management may often be the binding 'factor of production'. This suggests a focus on extension, training and education, perhaps with inducements to adopt IPM.

ADOPTING A PESTICIDE TAX

The earlier analysis suggested that a tax on pesticides could help move the system from the current excessive use of pesticides to one where pesticide use is substantially reduced. Taxes have the advantage that they do not imply that damage from pesticides is infinite, which is effectively what outright bans do. The problems of devising pesticide taxes are, however, quite significant:

- The chemicals involved are changing frequently – new pesticides are continually introduced and old ones phased out. This means that data go out of date relatively quickly.
- Many of the damage effects, in particular the chronic health impacts, occur through small doses over long time spans. So it is often impossible to quantify these effects directly, especially given that the chemicals involved are changing over time. Instead, tests using laboratory animals given relatively high doses over shorter periods must be used. The results of such tests may give only a rough indication of the likely effects in humans.
- The principal routes of non-work-related human exposure are through food contaminated with pesticide residues and through contaminated water. The levels of contamination may be very low, often undetectable with existing tests. For example, EPA food testing is capable of detecting only half the pesticide compounds of which residues might be present. This does not necessarily mean that the contamination is insignificant, because some compounds can accumulate in the body, and some could have an effect even at very low concentrations. But it does make it difficult to know what people are exposed to.
- Estimation of the actual exposure of different individuals is made even more difficult by the dependency of exposures on the specific items of food and water consumed by each individual. This hampers any attempt to relate pesticide exposure to illness in a population.

- For some pesticides, there may be complex synergistic effects with other man-made or natural chemicals. More generally, a variety of weather conditions, soil characteristics and so on can influence the damage effects of pesticides. The need to take these variables into consideration in the dose-response functions makes the estimation process more difficult.
- More generally, pesticides are non-point source pollutants and hence difficult to control if the aim is to set a tax proportional to damage done. Taxing pesticides at source is therefore a second-best solution but a workable one. However, Young et al (1994) suggest that diffuse sources are not an adequate reason to abandon the adoption of economic instruments. They argue for the adoption of pollution load permits allocated to farmers and based on a total pollution reduction target and various formulae for allocating it.

Nonetheless, several countries already tax the use of pesticides. Finland imposes a tax of 2.5 per cent on the price of pesticides – in other words, an *ad valorem* tax. Norway has a tax of 11 per cent on price. Sweden uses a more complicated system, involving a 20 per cent 'price regulation charge' on top of price, plus a 10 per cent 'environmental tax' on price and a charge equivalent to $5.65 per hectare per application. The charges in Norway and Finland are a step in the right direction. They advocate reduced use of pesticides and encourage farmers to look for alternative means of crop protection. However, they are not related to the potential damage caused by any given pesticide. They are not even based on the amount of active ingredient applied. Rather, the tax is applied to the purchase price of the pesticide, which may be completely unrelated to the ability of the pesticide to cause environmental damage. Therefore, perverse incentives might exist if the cheaper pesticides which farmers are encouraged to use are more environmentally damaging. The Swedish system also includes taxes based on the price of the pesticide, so the above comments apply. In addition, however, there is a charge on each application. This has the effect of discouraging any given application of a pesticide. While the charge is still far from perfect, it may be a little better related to the damage potential if it is thought that some damages, for example the risk of groundwater contamination, do not vary so much with the amount of chemical applied as with the frequency of applications.

While *ad valorem* taxes are usually not ranked as high as damage-related taxes in terms of social desirability, some authors suggest that they are superior in efficiency terms to runoff incentives – where the tax would be proportional to the context of pesticides in runoff – to pesticide standards (limits of pesticides in runoff), and to management practice standards (laying down restrictions on the number of pesticide applications); see Shortle and Dunn (1986) and Shumway and Chesser (1994). Shumway and Chesser simulate a 25 per cent *ad valorem* tax in south-central Texas, where groundwater quality is a sensitive issue. Most crops decline in terms of output, from 10 per cent (sorghum) to 50 per cent (corn), and the rate of decline in water quality was predicted to improve dramatically.

Various US states apply taxes to pesticides, mainly through registration fees, to generate funds for monitoring or groundwater protection. The Center for Science in the Public Interest (1995) has suggested the extension of systems already operating in several states where modest fees are charged in order to generate funds for public and occupational information programmes. The center's own opinion poll suggested that 84 per cent of respondents would be willing to pay $2 per year per family member for groceries that are free of pesticides. Robinson et al (1995) offer helpful formulae for devising better pesticide taxes.

- The 'mill' tax: California has taxed pesticides since 1971 when the tax was set at six mills ($0.006) per dollar on the wholesale price of the chemical. In 1995 the tax was 22 mills ($0.022); however, because the current tax includes some temporary items, the base rate of nine mills will operate from 1997. The tax yield appears to be about $25 million per annum. One simple tax solution, then, would be to introduce mill taxes more widely and to increase their size. The problem with such simple taxes is that they are proportionate to price and do not therefore discriminate between pesticides according to their toxicity, as would be required by the polluter pays principle. In other words, taxes should be at least approximately proportional to damage done.
- A tax could be proportionate to the active ingredients (AIs) in the pesticide. This has the attraction that the tax would be related in part to the risks associated with the pesticide, since risk will be related to the proportion of active ingredients. But one effect would be to discriminate against pesticides with low prices per pound of AI – and some of these are relatively non-toxic, as is the case with sodium chlorate. Risk-related taxes also have other difficulties in that risks vary across receptor: some have high toxicity but may rank low on groundwater contamination or ecological impacts.
- A tax could be varied according to the 'sensitivity' of the area in which it is applied. Highly biodiverse but vulnerable areas, for example, would attract a higher tax on pesticides than less diverse, more resilient areas. The obvious problems with such a tax would be the temptation to buy pesticides in low tax areas and import them into high risk areas. Other difficulties concern the complexity of ranking areas by vulnerability characteristics.

Stabinsky et al (1994) suggest a package of measures to induce farmers to make conversions to reduced pesticide packages, including cost sharing with government and tax credits. Clearly, then, damage-related taxes are likely to be very difficult to implement in practice, suggesting that simpler, though far less discriminating taxes may be best. To what extent would such taxes reduce the demand and use of pesticides ? This depends on the *elasticity of demand*, a measure of the responsiveness of demand to price. Estimates of the price elasticity of demand in the US put the figure at 0.1 to 0.5, which means that a 10 per cent increase in price due to the tax would reduce demand by 1 to 5 per cent. To secure major reductions in use, say 50 per cent, then, would require punitive taxes of 100 to 1000 per cent. But while the elasticities tend to be low, the positive side of the picture is that even modest taxes will raise quite large revenues. Robinson et al (1995) estimate that a doubling of the California mill tax rate, for example, would raise an extra $22 million per annum, producing a sizeable additional fund to finance education and information programmes.

Overall, devising pesticide taxes that discriminate by risk is likely to prove exceedingly difficult. This suggests deviating from the polluter pays principle in so far as a simpler tax base, such as value, should be used. Elasticities of demand do not suggest that feasible taxes will have a major effect on demand, but revenue generation should be high and provided such taxes are 'earmarked' for information and other pesticide control purposes, they should work well.

CONCLUSIONS

This chapter has reviewed the available literature on the costs and benefits of pesticide use. The main conclusion is that the literature offers some guidance only on the quantitative extent to which excessive environmental damage costs are imposed on society through the use of pesticides. There are several reasons for this.

- The most widely cited studies, those of Pimentel and colleagues (1991, 1993), seek to estimate the total damages from pesticide use, rather than the marginal damages done by additional pesticide applications, or the 'non-optimal damages'; they have methodological flaws, such as relying on costs of clean up rather than estimates of willingness to pay to avoid damage. Furthermore, they fuse damage done by pesticides with damages done by the misuse of pesticides. The last of these is important, because the policy implications may be quite different: while any form of excessive external damage implies that current practices should be altered, misuse may be better addressed through improved information and enforcement of safe use practice, rather than by quantity reductions as such. The only update of the Pimentel work which corrects some of the methodological errors, by Steiner et al (1995), suggests significantly lower US external costs from pesticide use. Again, the implications for policy are unclear because what is measured is a total external cost and the resulting benefit-cost ratios do not provide guidance on the extent to which pesticide consumption should be reduced.
- Other studies tend to be piecemeal, denying the possibility of generalizing from them. Global fatalities and poisoning estimates are at best good guesses, since the required information is not collected in a systematic way. Going by some of the estimates of fatalities, the monetary value of health damage could be substantial. This conclusion would be reinforced if the suggestion that pesticides are implicated in immune deficiency is found to be correct (Repetto and Baliga, 1996). However, most studies relate to total damages and give no indication of the relationship between the amount of pesticides used and the number of deaths. The work of Pingali et al (1995) is important in the developing country context in showing that the benefits from pesticide use may actually be outweighed by the health impacts, suggesting the need for much stronger controls on misuse, and much stronger incentives for alternative pesticide 'packages' of the form used in integrated pest management (IPM).
- Benefit-cost studies of IPM appear to suggest that it has lower overall costs than conventional pesticide use and that profits are positive even after allowing for yield effects. The issue, then, is why IPM has not captured a wider share of the market, and the literature indicates that IPM places demands on managerial resources and learning processes. If so, policy needs to be aimed at overcoming those barriers to adopting IPM.
- Micro studies, which estimate the willingness to pay of the public to avoid pesticide damages (which is the correct measure of damages), are comparatively few in number. Nonetheless, they suggest that consumers in developed economies are willing to pay premia for low or zero pesticide-residue produce. They may not be prepared to accept blemished produce just because it is pesticide free. The challenge, then, is both to reduce residues and maintain product quality in terms of cosmetic appearance. Just as importantly, the work of van

> Ravenswaay and colleagues shows the importance of acting on the 'uncertainty of risk': uncertainty about probabilities of risk and the need to act on forms of assurance that regulations are actually applied and enforced.

What, then, can be said about the economics of pesticide use? It is clear that there are external costs to pesticide use in the form of ecological and health damage. It is also clear that some of these damages relate to poor management of pesticides and their application, rather than to pesticide use. Finally, it is also fair to conclude that these damages are generally not internalized into farmers' application decisions. But the scale of the damages relative to the benefits of pesticide use remains uncertain, and it is this relative assessment which is needed before appropriate policy measures can be determined.

Policy measures that are currently justified would appear to be of three kinds. Firstly, there is a need to combat misuse of pesticides through better information campaigns and farmer education, especially in developing countries. Such measures might be associated with the gradual removal of remaining subsidies so that prices better indicate that the resource in question needs to be used economically. Secondly, where externalities do appear to be significant, a pesticide tax is an appropriate response, following the examples of those countries which have introduced such taxes: Finland, Norway, and Sweden, for example. Many economists would argue that an ideal pesticide tax would not apply uniformly to all pesticides – as in these countries – but would, rather, vary according to the best available assessment of the environmental damage potential of each pesticide under standard application conditions. Others suggest that *ad valorem* taxes may, in fact, be good approximators of a desirable tax.

Thirdly, it is possible to get some idea of the externalities associated with pesticides by approaching the issue from the side of consumer demand. Consumers appear willing to pay significant premia for reduced pesticide residues in produce, and this can be taken as a first order approximation of the externalities they perceive as resulting from produce. Such figures can be used to indicate the extent to which higher costs in pesticide-reduced production can go, the producer being able to pass those costs on to consumers as acceptable price increases to pay for the safety benefit. Various issues remain outstanding, however. The willingness-to-pay studies focus on the personal safety of the consumer, not the environmental damage or implications for health damage among workers. Nor are risk perceptions the same as 'objective risk': if consumers perceive the risk as being one in a thousand when in fact it is one in a million, what should policy-makers do ? If these differential risk assessments remain after information campaigns have been mounted, then there is a familiar problem present in all risk contexts, as some countries have found with, for example, the salmonella scare and BSE or CJD scares: failure to appreciate and act on perceived risk can lead to consumer vengeance. They simply cease to buy the product.

References

Angehrn B (1996) *Plant Protection Agents in Developing Country Agriculture: empirical evidence and methodological aspects of productivity and user safety*. Ph.D dissertation, Swiss Federal Institute of Technology, Zürich.

Antle JM and Pingali PL (1994) 'Pesticides, productivity, and farmer health: a Philippine case study'. *American Journal of Agricultural Economics* 76: pp 418–30. Also reprinted in Pingali et al (1995).

Ayer H and Conklin N (1990) 'Economics of ag chemicals: flawed methodology and a conflict of interest quagmire'. *Choices* Fourth Quarter 1990 pp 24–30.

Baker GA and Crosbie PJ (1993) 'Measuring food safety preferences: identifying consumer segments'. *Journal of Agricultural and Resource Economics* 18:2 pp 277–87.

Beach ED and Carlson GA (1993) 'A hedonic analysis of herbicides: do user safety and water quality matter?' *American Journal of Agricultural Economics* 75 pp 612–623.

Bowles R and Webster J (1995) 'Some problems associated with the analysis of the costs and benefits of pesticides'. *Crop Protection*, 14(7) pp 593–600.

Brouwer FM, Terluin IJ and Godeschalk FE (1994) *Pesticides in the EC.* Agricultural Economics Research Institute, The Hague.

Buttel F (1993) 'Socioeconomic impacts and social implications of reducing pesticide and agricultural chemical use in the United States' in D.Pimental and H.Lehman (eds) *The Pesticide Question: Environment, Economics and Ethics*, pp 153–181. Chapman and Hall, London.

Buzby J and Skees J (1994) 'Consumers want reduced exposure to pesticides on food'. *Food Review* May–August, pp 19–22.

Buzby J, Skees J and Ready R (1995) 'Using contingent valuation to value food safety: a case study of grapefruit and pesticide residues', in Caswell (ed) *Valuing Food Safety and Nutrition*, pp 219–56. Westview Press, Boulder, Colorado.

California Department of Pesticide Regulation (1993) *Summary of Illnesses and Injuries Reported by California Physicians as Potentially Related to Pesticides.* (Publication location not stated).

California Department of Food and Agriculture (1996) *Methyl Bromide: an Impact Assessment.* (Publication location not stated).

Campbell H (1978) 'Estimating the marginal productivity of agricultural pesticides: the case of treefruit farms in the Okanagan Valley'. *Canadian Journal of Agricultural Economics*, 24(2), pp 23–30.

Carlson G and Wetzstein M (1993) 'Pesticides and Pest Management', in Carlson G, Silberman D and Miranowski JA (eds) *Agricultural and Environmental Resource Economics.* Oxford University Press, New York.

Carlson GA (1977) 'Long-run productivity of insecticides'. *American Journal of Agricultural Economics*, 59(3) pp 543–48.

Caswell M (ed) (1995) *Valuing Food Safety and Nutrition.* Westview Press, Boulder, Colorado.

Center for Science in the Public Interest (1995) *Funding Safer Farming: Taxing Pesticides and Fertilisers.* Center for Science in the Public Interest, Washington, DC.

Chambers RG and Lichtenberg E (1994) 'Simple econometrics of pesticide productivity'. *American Journal of Agricultural Economics* 76, pp 407–17.

Chemishanksa L (1982) 'Long-term effects of pesticides: toxicological hazards, human health' in *Toxicology of Pesticides: Proceedings of a Seminar.* Sofia, Bulgaria, 31 August–12 September pp 54–66, World Health Organization.

Crosson P (1995) *An Income and Product Account Perspective of the Sustainability of US Agriculture*, Discussion Paper 95–27. Resources for the Future, Washington, DC.

Edwards S (1988) 'Option prices for groundwater protection'. *Journal of Environmental Economics and Management*, 15, pp 475–87.

Eom Y (1994) 'Pesticide residue risk and food safety valuation: a random utility approach'. *American Journal of Agricultural Economics*, November, pp 760–71.

Faeth P and Westra J (1993) 'Alternatives to corn and soybean production in two regions of the United States', in Faeth P (ed) *Agricultural Policy and Sustainability: Case Studies from India, Chile, the Philippines and the United States*, pp 63–92. World Resources Institute, Washington, DC.

Farah J (1994) *Pesticide Policies in Developing Countries: Do They Encourage Excessive Use?* World Bank Discussion Papers No 238. World Bank, Washington, DC.

Fischer LA (1970) 'The economics of pest control in Canadian apple production'. *Canadian Journal of Agricultural Economics*, 18(3) pp 89–96.

Freeman A (1993) *The Measurement of Environmental and Resource Values.* Resources for the Future, Washington, DC.

Gianessi LP (1993) *Issues Associated with the Benefits Assessment of Pesticides.* National Center for Food and Agricultural Policy, Washington, DC, December

Gianessi P (1991) *Reducing Pesticide Use with No Loss in Yields? A Critique of the Recent Cornell Report*, Discussion Paper QE 91–16. Resources for the Future, Washington, DC.

Georghiou G (1986) 'The magnitude of the resistance problem' in National Research Council, *Pesticides Resistance: Strategies and Tactics for Management.* National Academy Press, Washington, DC.

Headley JC (1968) 'Estimating the productivity of agricultural pesticides'. *American Journal of Agricultural Economics*, 50.

Higley L and Wintersteen W (1992) 'A novel approach to environmental risk assessment of pesticides as a basis for incorporating environmental costs into Economic Injury Levels'. *American Entomologist* 38 (1) pp 34–9.

Horowitz J (1994) 'Preferences for pesticide regulation'. *American Journal of Agricultural Economics* 76: pp 394–406.

Jeyaratnam J (1990) 'Acute pesticide poisonings: a major global health problem'. *World Health Statistics Quarterly*, 43, pp 139–43.

Kishor NM (1992) *Pesticide Externalities, Comparative Advantage, and Commodity Trade: Cotton in Andhra Pradesh, India*, World Bank Working Papers WPS 928.

Kishor NM (1993) 'The Effects of Pesticide Externalities on Cotton Cultivation in Andhra Pradesh'. mimeo

Knutson RD, Taylor CR, Penson J and Smith E (1990) *Economic Impacts of Reduced Chemical Use.* Knutson and Associates, College Station, Texas.

Lee L (1992) 'A perspective on the economic impacts of reducing agricultural chemical use'. *American Journal of Alternative Agriculture*, 7(1–2), pp 82–88.

Moore C and Villarejo D (1996) *A Critique of the Report: Economic Impact of Methyl Bromide Cancellation.* California Institute for Rural Studies, Davis, CA.

Natural Resources Defense Council (1992) *Intolerable Risk: Pesticides in Our Children's Food.* NRDC, Washington, DC.

Norton G and Mullen J (1994) *Economic Evaluation of Integrated Pest Management Programs: a Literature Review*, Publication 448–120, Virginia Cooperative Extension, Blacksburg, VA.

Ott S (1990) 'Supermarket shoppers' pesticide concerns and willingness to purchase certified pesticide residue-free produce'. *Agribusiness: An International Journal* 6(6) pp 593–602.

Pearce DW, Whittington D and Georgiou S (1994) *Project and Policy Appraisal, Integrating Economics and the Environment*, OECD, Paris.

Pearce DW, Bann C and Georgiou S (1992) *The Social Cost of Fuel Cycles.* HMSO, London.

Pimbert MP (1991) *Designing Integrated Pest Management for Sustainable and Productive Futures.* International Institute of Environment and Development (IIED), Gatekeeper Series No 29.

Pimentel D, Andow D, Dyson-Hudson R, Jacobson S, Irish M, Kroop M, Moss A, Schreiner I, Shepard M, Thompson T and Vinzani B (1980) 'Environmental and social costs of pesticides: a preliminary assessment'. *Oikos*, 34 (2), pp 126–40.

Pimentel D, McLaughlin L, Zepp A, Lakitan B, Kraus T, Kleinman P, Vancini F, Roach W, Graap E, Keeton W, Selig G (1991a) 'Environmental and economic impacts of reducing agricultural pesticide use'. *Bioscience*, 41 (6), pp 403–409.

Pimentel D, McLaughlin L, Zepp A, Lakitan B, Kraus T, Kleinman P, Vancini F, Roach WJ, Graap E, Keeton WS and Selig G (1991b) 'Environmental and economic impacts of reducing US agricultural pesticide use' in Pimental D and Lehman H (ed) *The Pesticide Question: Environmental, Economics and Ethics*, pp 223–80. Chapman and Hall, London.

Pimentel D, McLaughlin L, Zepp A, Lakitan B, Kraus T, Kleinman P, Vancini F, Roach WJ, Graap E, Keeton WS and Selig G (1991c) 'Environmental and economic impacts of reducing US agricultural pesticide use' in Pimental D (ed) *Handbook of Pest Management in Agriculture*, 2nd ed. CRC Press, Boca Raton, FL.

Pimentel D, Acguay H, Biltonen M, Rice P, Silva M, Nelson J, Lipner V, Giordano S, Harowitz A and D'Amore M (1992) 'Environmental and economic costs of pesticide use'. *Bioscience*, 42(10), pp 750–60.

Pingali P, Marquez C and Palis F (1994) 'Pesticides and Philippine rice farmer health: a medical and economic analysis'. *American Journal of Agricultural Economics* 76, pp 587–92.

Pingali P, Marquez C, Palis F and Rola, A. (1995) 'The impact of long term pesticide exposure on farmer health: a medical and economic analysis of the Philippines' in Pingali, P and Roger, P (1995) *Impact of Pesticides on Farmer Health and the Rice Environment*, Kluwer, Dordrecht, The Netherlands, pp 343–60.

Pingali P and Roger P (1995) *Impact of Pesticides on Farmer Health and the Rice Environment*. Kluwer, Dordrecht, The Netherlands.

Pretty J (1993) *Sustainable Agriculture in Britain: Recent Achievements and New Policy Challenges*, Research Series 2(1). International Institute for Environment and Development, London.

Pretty J (1995) *Regenerating Agriculture*, Earthscan, London.

Rands MRW and Sotherton NW (1986) 'Pesticide use on cereal crops and changes in the abundance of butterflies on arable farmland in England'. *Biological Conservation*, 36: pp 71–82.

Rendleman CM (1991) 'Agrichemical reduction policy: its effects on income and income distribution'. *Journal of Agricultural Economics Research* 43(4), pp 3–9.

Repetto R (1985) *Paying the Price: Pesticide Subsidies in the Developing Countries*. World Resources Institute, Washington, DC.

Repetto, R (1989) 'Economic Incentives for Sustainable Production' in Schramm, G and Warford, J (eds) *Environmental Management and Economic Development*. Johns Hopkins University Press, Baltimore.

Repetto, R and Baliga, S (1996) *Pesticides and the Immune System: the Public Health Risks*. World Resources Institute, Washington, DC.

Reus JAWA, Weckseler HJ and Pak GA (1994) *Towards a Future EC Pesticide Policy*, report to European Community. Centre for Agriculture and Environment (CLM), Utrecht, The Netherlands.

Robinson J, Tuden D and Pease S (1995) *Taxing Pesticides to Fund Environmental Protection and Integrated Pest Management*. Center for Occupational and Environmental Health, University of California, Berkeley.

Rola A and Pingali P (1993) 'Pesticides, rice productivity and health impacts in the Philippines'. in Faeth P (ed) *Agricultural Policy and Sustainability: Case Studies from India, Chile, the Philippines and the United States*. World Resources Institute, Washington, DC.

Schmitz, PM and Hartmann M (1993) *Landwirtschaft und Chemie*. Wissenschaftsverlag Vauk, Kiel.

Shortle J and Dunn J (1986) 'The relative efficiency of agricultural source water pollution control policies'. *American Journal of Agricultural Economics*, 68, pp 668–77.

Shumway C and Chesser R (1994) 'Pesticide tax, cropping patterns, and water quality in south central Texas'. *Journal of Agricultural and Applied Economics*, 26(1), pp 224–40.

Smith VK (1992) 'Environmental costing for agriculture: will it be standard fare in the Farm Bill of 2000?' *American Journal of Agricultural Economics*

Stabinski D, Moore M, Jennings B, and Rosenthal E (1994) *Financial Incentives and their Potential to Reduce Pesticide Use in Three California Crops.* Pesticide Action Network, San Francisco, CA.

Steiner R, McLaughlin L, Faeth P and Janke R (1995) 'Incorporating externality costs into productivity measures: a case study using US agriculture' in Barbett V and Payne R and Steiner R (eds) *Agricultural Sustainability: Environmental and Statistical Considerations.* John Wiley & Sons, New York, pp 209–30.

Thrupp LA (1990) 'Entrapment and escape from fruitless insecticide use: lessons from the banana sector of Costa Rica'. *International Journal of Environmental Studies* 36, pp 173–89.

US Department of Agriculture (1994) *Agricultural Resources and Environmental Indicators.* Agricultural Handbook No 705. USDA, Washington, DC.

US Environmental Protection Agency (EPA) (1997) *Pesticide Industry Sales and Usage: 1994 and 1995 Market estimates.* US EPA, Washington, DC.

US Office of Technology Assessment (1995) *Targeting Environmental Priorities in Agriculture: Reforming Program Strategies.* OTA-ENV–640, Government Printing Office, Washington, DC.

van Ravenswaay E and Hoehn J (1991a) 'The impact of health risk information on food demand: a case study of Alar and apples' in Caswell, JA (ed) *Economics of Food Safety.* Elsevier, New York.

van Ravenswaay E and Hoehn J (1991b) *Consumer Willingness to Pay for Reducing Pesticide Residues in Food: Results of a Nationwide Survey,* Paper 91–18. Department of Agricultural Economics, Michigan State University.

van Ravenswaay E and Hoehn J (1991c) *Consumer Perspectives on Food Safety Issues: The Case of Pesticide Residues in Fresh Produce,* Paper 91–20. Department of Agricultural Economics, Michigan State University.

van Ravenswaay E and Hoehn J (1991d) *Contingent Valuation and Food Safety: The Case of Pesticide Residues in Food,* Paper 91–13. Department of Agricultural Economics, Michigan State University.

van Ravenswaay E, Wohl J and Hoehn J (1992) *Michigan Consumers' Perceptions of Pesticide Residues in Food,* Paper 92–56. Department of Agricultural Economics, Michigan State University.

van Ravenswaay E and Wohl J (1995) 'Using contingent valuation methods to value the health risks from pesticide residues when risks are ambiguous' in Caswell J (ed) *Valuing Food Safety and Nutrition.* Westview Press, Boulder, CO.

Warburton H, Palis F and Pingali P (1995) 'Farmers' perceptions, knowledge and pesticide use practices' in Pingali P and Roger P (eds) *Impact of Pesticides on Farmer Health and the Rice Environment.* Kluwer, Dordrecht, pp 59–96.

Weaver RD, Evans J and Luloff E (1992) 'Pesticide use in tomato production: consumer concerns and willingness to pay'. *Agribusiness: an International Journal,* 8, pp 131–42.

WHO (1990) *Public Health Impact of Pesticides in Agriculture.* World Health Organization, Geneva.

Wohl J, Ravensway E and Hoehn J (1995) *The Effect of Ambiguity on Willingness to Pay for Reduced Pesticide Usage,* Paper 95–41. Department of Agricultural Economics, Michigan State University.

WRI (World Resources Institute) (1994) *World Resources 1994–1995.* Oxford University Press, Oxford.

Young T and Congdon C (1994) *Plowing New Ground: Using Economic Incentives to Control Water Pollution from Agriculture.* Environmental Defense Fund, Oakland, CA.

Zilberman D, Campbell M and Manale A (1994) 'Who Makes Pesticide Use Decisions: Implications for Policy Makers'. *Proceedings of a Workshop on Pesticide Use and Produce Quality,* Farm Foundation, Illinois.

Zilberman D, Schmitz A, Casterline G, Lichtenberg E and Siebert J (1991) 'The economics of pesticide use and regulation'. *Science,* 253 (5019), pp 518–23.

4

Incorporating External Impacts in Pest Management Choices

Vivien Foster, Susana Mourato, Robert Tinch, Ece Özdemiroğlu and David Pearce

Pesticide Ratings and Contingent Ranking of Effects

The previous chapter covered in some detail the many reasons for supposing that the levels of pesticide use chosen by growers may be too high from the perspective of society as a whole. The concept of the 'true price' of pesticides was introduced as the sum of the market price and the marginal external effects. Analysis was presented showing that a socially optimal outcome requires the marginal benefits of pesticide use to be equal to the marginal social costs, that is, the true price. The chapter continued to examine some attempts which have been made to quantify, firstly, the benefits of pesticide use and, secondly, the external costs. The various shortcomings of these attempts were noted, in particular the tendency to focus on total rather than marginal changes. A combination of theoretical arguments and empirical evidence was used to show, on a general scale, that current levels of pesticide use are likely to be too high.

How can these insights be incorporated within pest-control decisions? Firstly, we need to consider what these decisions are. A little thought shows that matters do not rest at the total quantity of pesticides used. There is also the type of pesticides, and the timing, frequency and technique of application. Going back a stage further, we could consider also the choice of crops and the locations in which they are grown. The important point is that all of these aspects of the decision process could be suboptimal from the social standpoint for the same basic reasons as discussed in the previous chapter.

The economist's ideal scenario is a full and accurate assessment of the marginal benefits and costs of pesticide use in *any given situation*, that is, for each aspect of the decision process faced by each grower. However, this is likely to be very difficult, if not impossible, to achieve. Alongside the complexity detailed in the previous paragraph, one of the principal reasons for this is that the specifics of the situation vary greatly from case to case: different crops, pesticide characteristics, climatic conditions, arthropod communities, proximity to watercourses – the list is potentially endless. Add to this the considerable scientific uncertainties about both the relationships between different links in the system and the risks of damage, and it becomes clear that a workable solution will involve many simplifying assumptions. If it is not possible fully to quantify the costs and benefits for each situation, how then are actual decisions to be moved closer to theoretical optima? Several innovative techniques have been developed to make best use of the information available, including pesticide-ranking indices, surveys of farmers' preferences and surveys of consumer preferences. The rest of this chapter discusses these techniques and how they can be used for policy-making.

Pesticide-ranking indices

In recent years, a small number of studies have taken the approach of trying to rank pesticides in terms of their potential to cause external damages. One of the earliest is that of Kovach et al (1992). They attempt to convert information on the toxicological and environmental characteristics of pesticides into a form more readily interpretable, and directly useful for decision makers, which they term the environmental impact quotient (EIQ). Considerable simplification is involved in calculating EIQ. It is a composite index formed by arithmetic combinations of several other indices, each of which is based on considering the characteristics of the pesticides. For example, one component of the EIQ is farmworker risk, which itself is formed of 'applicator effects' and 'picker effects', which depend on three-point scales for dermal toxicity, chronic toxicity and, for picker effects, plant surface half-life. The other parts of the EIQ similarly trace back to three-point scales for various objectively measurable characteristics.

The use of three-point scales – effectively, high, medium, and low risk – is clearly a highly simplifying assumption. Nevertheless, this in itself is not necessarily objectionable, given the significant uncertainties involved. Indeed, it is likely that the rankings within any given category are reasonably reliable. That is to say, the index for bird effects probably gives a fair representation of the relative potentials of different pesticides to cause damage to birds. This is also the case for bee effects, and so on. However, a problem arises in trying to combine these indices to form the overall EIQ. There is no way to do this objectively on the basis of the scientific data, because any combination of the indices involves (whether explicitly or implicitly) weighting of the relative values of damage to different categories.

How important is this criticism? One way to check is to consider the rankings within each category of effect which are calculated by Kovach et al for different pesticides. If pesticides which are relatively more damaging on one scale, say aquatic effects, tend also to be relatively more damaging on other scales, then the criticism is of minor importance because the ranks will not change much however the individual components are combined. In fact, Kovach et al show that, while some of the indices display high correlations, others display little or no correlation, while some even show negative correlations. Groundwater effects, for example, show either no correlation or a weak negative correlation with all the other categories. This means that herbicides which are potentially more damaging to groundwater tend to be less damaging to some other categories. Although the correlation is usually weak, the implication is clear: if we consider groundwater effects to be of little importance, and weight them as such in our EIQ calculation, then the EIQs will rank herbicides in one way. If, however, we weight groundwater risks very heavily in our EIQ calculation, the EIQs will rank herbicides in a way that is the reverse of the first. This makes clear the importance of the weighting factors in calculating the EIQ. Rather than leave this weighting as a subjective decision, what is required is some way of determining objective weights for the different components.

Making rankings useful through objective weighting

This is where economics can be of assistance. In essence, the weights applied to each category should, in fact, be the relative 'values' which society places on the damages. So if we can measure these values, we will, with some simplifying assumptions, be able to derive relatively robust means of weighting the different categories to construct an overall EIQ.

The subject of environmental valuation has been touched on in the previous chapter. To recap, the basic value judgement involved is that individuals' preferences should form the basis of social decisions. For marketed goods, these preferences are expressed in demand behaviour through the willingness to pay (WTP) of individuals for different levels of different goods. Environmental valuation attempts to estimate the WTP of individuals for goods and services which people value, even though they are not bought and sold on a market. This may be achieved by revealed preference, which looks at certain aspects of actual behaviour, or stated preference, which elicits decisions from individuals faced with hypothetical situations.

Relative and absolute values

It is worth noting that it is only necessary to measure relative values in order to determine weights for constructing composite indices. That is to say, it is sufficient to estimate the ratio at which different components are traded off – that is, the ratio of their WTPs. It is not strictly necessary to know the absolute value of the WTPs. This is potentially important with respect to stated preference techniques.

These techniques are sometimes criticized on the grounds that people state higher WTP in hypothetical situations than they would actually pay in reality. Contingent ranking is a stated preference technique which analyses trade-off decisions between several different characteristics, including money. It is possible that the money valuations implied by use of this technique may be too high in some cases. However, the *relative* WTPs arising from contingent ranking, that is, the rate at which different environmental goods are traded off, is much less likely to be misrepresented. This is supported by evidence from market research, which makes extensive use of contingent ranking. A common finding is that relative values for different product characteristics correspond closely with reality. This can be expected to carry over to environmental valuation using contingent ranking.

Determination of relative weights would put us one step closer to the goal of socially optimal pesticide use. It would allow pesticides to be ranked objectively in terms of their potential to cause environmental and health damage. In situations where different pesticides can be applied with equal efficacy in terms of crop yield but with different environmental impacts, this ranking will be sufficient to determine which is the best choice.

Contingent valuation of farmers' preferences

An approach which, in many ways, followed this line of thought was taken by Higley and Wintersteen (1992). They too used scientific data to rank pesticides as high, mid, low and negligible risk with respect to eight health and environmental factors: surface water, groundwater, aquatic organisms, birds, mammals, beneficial insects, acute human toxicity, and chronic human toxicity.

A survey of northern US field-crop growers was conducted with two aims: to determine how to weight the categories and to determine their WTP to avoid different levels of risk. Firstly, farmers were asked for their ratings, on a scale of one to ten, of the importance of each damage category. Mammal and bird toxicity were the two lowest categories, while human chronic and acute toxicity and groundwater contamination came out highest: but all fell within a fairly narrow range. These responses were used to calculate a relative importance rating for each category, which was the mean rating for that category divided by the total means of all categories. Secondly, a simple series of contingent valuation questions was used to derive farmers' WTP for avoiding high, mid and low risk from a single pesticide application per acre. The best mean estimates from the study gave values of $12.54, $8.76 and $5.79 per acre respectively. These results were used to calculate the environmental costs per acre per application of some pesticides. This is done by multiplying the relative importance rating by the risk value ($12.5, $8.8, $5.8, or zero) for each damage category, and summing over all eight categories. The overall value ranges potentially from zero for pesticides having no risk of any category of damage to $12.54 for pesticides scoring high risk for all categories. These scores are termed economic injury levels (EILs).

One shortcoming with this approach is that the WTP values, and indeed the rankings of different categories, reflect only farmers' preferences and not those of the rest of society. The authors note this, adding, however, that because the use of EILs is currently optional, it does not make sense to derive environmentally augmented EILs using preferences other than those of farmers themselves. In other words, Higley and Wintersteen aim to present environmentally-concerned growers and other pesticide decision-makers with scientific data in a user-friendly format – a useful goal in itself, but different from the objective of reaching socially optimal pesticide-use decisions.

Two important improvements could be made to the methodology. Firstly, a more robust way of rating the importance of different damage categories is required. As conducted by Higley and Wintersteen, the process involves independent scoring of the importance of each category, which does not really address the key issue of how damage to these categories is traded off in peoples' preferences. Secondly, it would be preferable to combine the importance rating and risk valuation parts of the exercise: in Higley and Wintersteen, the valuation was of high, mid or low risk, without specifying what exactly the risk was, nor to what categories it applied. The technique of contingent ranking will be advanced below as a means of satisfying both of these objectives.

Contingent-ranking estimates of pesticide external costs

All environmental valuation techniques, in particular stated preference techniques, are costly to conduct in both time and money. Of course, with an issue as ubiquitous and significant as pesticide use, these expenditures will be justified. But for the purposes of this book, only a fairly small-scale survey could be afforded. This is intended primarily to show how economic valuation techniques could be used to provide comprehensive external-cost estimates, although it is also hoped that the estimates may be of some interest in themselves.

Kovach et al (1992) use 16 basic characteristics of pesticides (such as dermal toxicity, soil half-life, etc) to derive eight primary categories of effect (such as applicator and groundwater effects). These are further combined to form three components compris-

Table 4.1 *Correlation between health and ecological risks for pesticides*

Pesticide group	*Insecticides*	*Fungicides/nematicides*	*Herbicides*
Health-ecology correlation	0.54	0.37	0.52

Source: Calculated from Kovach et al (1992)

ing the EIQ: farmworker component, consumer component and ecological component. Although to estimate relative WTP values for each of the eight impact categories would be of great potential value, it would have exceeded the capacity of a survey of this size. Overloading the respondents with more choices than they can meaningfully evaluate would have been counterproductive.

For the purposes of exposition and development of the methodology, Kovach et al focused on two main factors: health damages and ecological damages. A health risk index was given by the average of the consumer component and the farmworker component for each pesticide.

Table 4.1 shows the correlation coefficients between the health risk index and the ecological index for three different groups of pesticides. Not surprisingly, the coefficients are higher than those between some of the individual categories: these two indices are more inclusive, and have some basic data in common, both using water and plant surface half-lives, for example. Nonetheless, the correlations are weak, in particular for fungicides and nematicides, suggesting that cases will arise in which a pesticide preferable on the health scale will be unfavourable on the ecological scale.

Here, weights are generated by combining the health index with the ecological index. But the component parts of these indices have been combined in the same arbitrary way as in the original study. For example, the ecology component is derived from arbitrary combinations of indices for bird effects, bee toxicity, beneficial insect effects, and aquatic effects. More extensive contingent ranking exercises would be required to derive robust ways of combining these component parts, and so on back to the base scientific data from which all the indices are ultimately constructed.

CONTINGENT-RANKING EXERCISE

The contingent-ranking approach

Contingent ranking is a survey method which is designed to isolate the value of individual product *attributes* (or characteristics) which are typically supplied in combination with one another. This is achieved by constructing a number of hypothetical variations on a particular product, each with different *levels* of the relevant attributes, and asking respondents to rank these variations according to their personal preferences. One of these attributes must be price, but the other attributes may reflect any qualitative product characteristics which are deemed to be of interest. When analysed using an appropriate statistical model, the ranking data permit the identification of willingness to pay for changes in the level of each of the qualitative attributes included in the survey.

The contingent ranking method has already been applied successfully to a number of economic problems, both within the context of environmental economics and beyond (Beggs et al, 1981; Lareau and Rae, 1985). The method has certain advantages and disadvantages compared to the related, and more commonly used, contingent valuation method. In contingent ranking, the respondent does not have to state willingness to pay for an unfamiliar good, but is given the simpler task of producing a ranking of alternatives from which a willingness to pay can ultimately be inferred. Moreover, unlike most contingent valuation surveys, a contingent ranking study is not confined to valuing a single dimension of change. Less positively, a contingent ranking exercise can rapidly become prohibitively complex once respondents are asked to consider more than a few attributes at different levels. Thus, there are definite cognitive limits on the sorts of ranking exercises that can realistically be requested from respondents.

Contingent ranking can be made relevant to the problem of valuing the environmental impacts of pesticides in the following way. The environmental damages incurred during the cultivation of an agricultural product can be thought of as attributes of that product, even if they do not lead to any ultimate physical effects on the nature of the product offered to consumers. Product labelling can be (and indeed has been) used to introduce a differentiation between products purely on the basis of their production processes. For example, through labelling, consumers have been able to view 'dolphin-friendly' tuna differently from tuna harvested by conventional methods, even though the physical characteristics of the two products on the supermarket shelf are indistinguishable. Following similar reasoning, the environmental damages associated with pesticides could be valued by asking people to rank a particular agricultural product according to the varying environmental impacts of its different production processes. This is the approach developed here.

The design of the survey

The design of a contingent-ranking survey entails a number of distinct stages: first, the focus product is identified; second, the relevant attributes are selected; third, the levels of those attributes to be presented to consumers are established; finally, the sets of hypothetical products to be ranked by each consumer are constructed. For a variety of reasons, bread was chosen as the consumer product on which the survey would be based. Arable crops account for some 90 per cent of pesticide usage in the UK, whether measured by area treated or weight applied (DoE, 1996). Wheat is also one of the major arable crops and probably the most familiar to consumers. Furthermore, all known economic valuation studies of pesticides to date have focused on fresh fruit and vegetables. With this category of produce there is the additional complication that reductions in pesticide usage lead to a noticeable deterioration in aesthetic quality. The aesthetic appearance of bread, on the other hand, is unaffected by marginal reductions in pesticide use. This simplifies the contingent-ranking problem so that attributes can be focused exclusively on the health and environmental impacts of pesticide use. In the survey, 'cases of pesticide poisonings among farmworkers and the general public each year attributable to wheat cultivation' counted as a *health attribute*, while the 'number of farmland bird species in a state of serious long-term decline' was adopted as an *ecological attribute* (For more details on the survey, see Foster and Mourato, 1997).

The survey provides an illustration of a methodology that can be used to obtain estimates for the social cost associated with pesticide impacts on the environment. This was based on the intuition that consumers are able to differentiate between goods according to the processes by which they are produced, provided that an adequate amount of information is supplied. The results suggest the following range of mean estimates of household willingness to pay values:

- 0.73 – 1.05 pence per loaf for a case of illness;
- 4.53 – 6.59 pence per loaf for a bird species in decline.

These are marginal values and should not be used to estimate total costs by multiplying the total effects: for example, people's WTP for reversing the decline in one bird species is not necessarily one ninth of their WTP for reversing the decline in all nine species. This type of insensitivity to scope is known as the 'embedding phenomenon'.

Although these results should be interpreted as an upper bound on the true valuation, any upward bias should not affect the *ratio* of willingness to pay between the health and environment categories: both WTP figures would be inflated by the same (unknown) factor. The results of the survey indicate that society's WTP to avoid a single case of mild human illness is very much lower than its WTP to reverse the decline in a species of farmland birds. The ratio of the two was 1:6.5.

The present study was necessarily experimental in character given the paucity of previous work on this issue. Ideally, future work should concentrate on introducing additional numbers of attributes or changing the nature of the attributes used. The cognitive limits, posed by respondents' ability to answer contingent-ranking questions, mean that this is likely to be a challenging task requiring quite a large-scale survey.

USE OF CONTINGENT-RANKING RESULTS IN PESTICIDE CHOICE DECISIONS

Calculating the overall marginal social cost of pesticides

An interesting policy question is the appropriate magnitude of a sales tax on pesticides designed to take into account the full marginal social cost. The contingent-ranking survey was not explicitly designed to answer this question; its aim was rather to uncover people's trade-offs between different categories of environmental damage in order to conduct weighting of the environmental and health impacts of different pesticides, as discussed below.

Nonetheless, subject to some caveats, the results can be used to illustrate how the level of a tax could be calculated from survey data. The correct level of the tax is given by the aggregate willingness to pay in terms of pence per loaf per unit of damage multiplied by the number of loaves produced per kilogramme of pesticide applied and by the total damage. To obtain the aggregate willingness to pay, the marginal valuation in terms of pence per loaf is multiplied by the total number of loaves purchased over a given time period, such as a week. The resulting figure is then divided by the total volume of pesticide applications to cereal crops over the same period.

The problem resides in aggregating the marginal willingness to pay estimate per unit of damage over the total level of damages. As was mentioned before, due to the broadly observed phenomenon of insensitivity to small quantity changes, the marginal values obtained in the survey should not be used to estimate total values by multiplying by the total damage level. A conservative approach is to assume that such embedding is complete and take the unit estimates as not differing significantly from the total value. Although the assumption that embedding is complete will almost certainly involve some downwards bias, this may be desirable for the purposes of estimating tax rates. This is because, as described in detail in the previous chapter, the theoretical optimal tax for a pollution source is equal to the marginal external costs *at the optimum*, where marginal external costs and marginal abatement costs are equal. There are good reasons to believe not only that marginal external costs increase with the amount of pesticides used, but also that they do so at an increasing rate. And this would suggest a conservative approach to estimating tax rates from current marginal cost figures.

To illustrate this calculation, the marginal WTP estimates from the 'most preferred alternative model' with the income interaction were chosen: 0.73 pence per loaf to avoid a case of illness and 4.53 pence per loaf to avoid losses in a bird species. Given the expected upward bias in the absolute WTP figures, it seems sensible to adopt the most conservative bids if the survey results are to be used for policy purposes. From the Family Expenditure Survey (Central Statistical Office, 1993), it can be inferred that, on average, a household in the UK consumes around 161 loaves of bread per year. Given that there are approximately 21 million households in the UK, the total number of loaves consumed each year is around 3381 million. According to Brouwer et al (1994), the total volume of pesticides used in cereal crops in the UK is approximately 15 million kilograms per year. Hence, marginal WTP in pounds per loaf * (total number of loaves/total volume of pesticide applications in kilogrammes) = estimated marginal external costs of pesticide use in £ per kilogramme.

$$(0.0073 + 0.0453) * 3381{,}000{,}000/15{,}000{,}000 = £12 \text{ per kilogramme}$$

Following this assumption, the resulting figure suggests an approximate 60 per cent uniform tax on pesticides (an estimated increase in average pesticide cost from an actual £20 to £32 per kilogramme). This lower-bound figure illustrates just how important the consideration of external effects is. It should, however, be interpreted with caution, given the tentative assumptions associated with the calculations and the already referred to problems associated with uniform taxes. So it should be taken more as an illustration of how the survey results can be used to calculate the external costs of pesticide application.

Ranking of selected pesticides used on wheat in the UK

One of the objectives of the contingent-ranking survey was to derive results which would be of general use in presenting a methodology for developing objective weightings for pesticide ranking exercises. The survey was conducted with reference specifically to the environmental impacts of wheat production in the UK, using a UK resident sample. This suggests that the actual WTP amounts derived would be specific to this situation. However, it is likely that the ratio of WTPs will be less situation-specific. Assuming that the relative risks to health and environmental damage are the same as those presented in the survey, it might be expected that the relative weightings of environmental and health damage would be similar were the study to be conducted in other countries, although the absolute value of WTP would certainly vary. This conclusion rests on the assumption that the structure of preferences is similar in different countries, even though income levels (and hence WTP) may vary substantially. It is probably a reasonably accurate assumption for transferring the results for use in other European and North American countries; it may be more questionable for use in developing countries where the differences are much greater.

According to Garthwaite et al (1994), wheat crops in the UK receive on average two fungicide applications (25 per cent of the total weight of active ingredients applied to wheat), two herbicides (43 per cent), one growth regulator (23 per cent) and one insecticide (5 per cent). Tables 4.2 to 4.4 present results comparing the health and environmental impacts of the various pesticides using the weights estimated by Kovach et al (1992) and the WTP results reported above. The comparisons are by no means

Table 4.2 *Comparison of insecticides used on wheat*

Insecticide	*Health*	*Ecology*	*EIQ*	*EIQ rank*	*New EIQ*	*New EIQ rank*
Chlorpyrifos	27	105	53	4	95	3
Dimethoate	40	141	74	5	128	5
Esfenvalerate	6	137	50	3	120	4
Fonofos	26	83	45	2	75	2
Pirimicarb	23	46	30	1	43	1

Note: Rank 1: the least damaging pesticide

exhaustive because the necessary primary physical data are not available for all pesticides used on wheat crops. Rather, the tables are meant to illustrate the idea of deriving objective ranks to aid pesticide choice, use and taxation decisions.

The ecology columns show the ecological component of Kovach et al's EIQ index, while the health columns are the average of their farmworker and consumer components. The EIQ column is the sum of their farmworker, consumer and ecological components, divided by three, as done by the authors. New EIQ columns show the ranking using the ratio of the WTP values as the relative weights which should be applied to each category. The contingent-ranking exercise produced a ratio of ecological to health WTP of 6.5. To calculate the modified EIQ values, therefore, the index of ecological risk has been weighted 6.5 times more heavily than the health index. The modified EIQ is therefore the sum of the health index and 6.5 times the ecological index, all divided by 7.5.

Tables 4.2 to 4.4 show how, by using different weighting measures, the implications of the EIQ might be changed. In a few cases, a pesticide which seems preferable to another under the old EIQ is found to be more damaging under the new EIQ, as is shown by the columns for EIQ rank and new EIQ rank. Consider, for example, chlorpyrifos and esfenvalerate in Table 4.2, both of which are used for the control of aphids and for general insect pest control. Under the EIQ, esfenvalerate seems slightly favourable, although there is little to choose between the chemicals. But under the new EIQ, chlorpyrifos seems substantially better, while esfenvalerate is seen as closer to the worst choice, dimethoate. Similarly, the herbicides glyphosate and MCPA (Table 4.4), both used for general weed control in wheat crops, trade places. Under the EIQ, MCPA seems slightly worse, while under the new, it is slightly better. More generally, the relative

Table 4.3 *Comparison of fungicides used on wheat*

Insecticide	*Health*	*Ecology*	*EIQ*	*EIQ rank*	*New EIQ*	*New EIQ rank*
Carboxin	8	45	20	1	40	1
Chlorothalonil	18	102	46	4=	91	4
Flusilazole	8	82	33	3	72	3
Iprodione	6	69	27	2	61	2
Mancozeb	28	130	62	6	116	6
Maneb	28	135	64	7	121	7
Sulphur	8	120	46	4=	105	5

Table 4.4 *Comparison of herbicides used on wheat*

Insecticide	*Health*	*Ecology*	*EIQ*	*EIQ rank*	*New EIQ*	*New EIQ rank*
Cyanazine	17	26	20	2	25	2
Glyphosate	12	74	32	5	66	6
MCPA	20	69	37	6	62	5
Paraquat	42	125	70	7	114	7
Pendimethalin	12	54	26	3	48	3
Simazine	10	26	16	1	24	1
Trifluralin	12	57	27	4	51	4

damage potential of pesticides as measured by the new EIQ are different from those under the old EIQ. This has implications for a differentiated tax policy, in that the tax rates for different chemicals will vary if the new EIQ is used.

In interpreting these results, it is important to remember that since the survey could look only at two components of the index – health and environment –changes in rankings from the EIQ system will be modest. A full attempt to derive objective weights for all the components of the index is likely to lead to rather more significant changes in the ranks – and correspondingly greater implications for design of a tax system.

In addition to the limited extent of the contingent-ranking survey, it has to be assumed that the relative risks to health and the environment in the survey are reasonably representative of the relative risks encapsulated in the health and ecological components of the EIQ index. This might not be true. Again, the intention is solely to advance a methodological approach. For any attempt to further refine the use of the EIQ, it will be necessary not only to conduct more extensive contingent-ranking studies, but also to study in more detail the relative levels of risk that apply in each category.

Using rankings to estimate a differentiated tax

It is fairly straightforward to use the above pesticide ranking tables to show what a differentiated tax would look like. The absolute tax levels can also be estimated if some additional simplifying assumptions are made. Rather than work through calculations for all the pesticides listed, we will focus on insecticides to illustrate the procedure. What is required of a differentiated tax is that the tax rate should be proportional to the index of environmental damage – in the present case, the new EIQ measure. Of course, the results will be dependent on the extent to which the new EIQ reflects accurately the relative damage potentials, so the caveats presented above apply. The column in Table 4.5 headed relative tax shows what the relative tax rates should be for the five insecticides shown, assuming a tax rate of £X per kilogramme for pirimicarb. For example, the table suggests that dimethoate should have a tax per kilogramme three times higher than that for pirimicarb.

To move to estimates of the actual tax rates which should be applied is more difficult. Based on the calculations presented in the previous section, we estimated an average tax of £12 per kilogramme. This average tax can be viewed as relating to the environmental and health impacts of an 'average' pesticide. Unfortunately, we have been unable to estimate a new EIQ figure for all pesticides used on cereal crops, due to shortage of scientific data on the properties of several pesticides.

To get round this problem, we will make the following rough estimate of the average new EIQ figures, as follows. The mean new EIQ figures from the above tables are approximately 92 for insecticides, 87 for fungicides and 56 for herbicides. This corresponds reasonably well with what might be expected regarding the relative potentials for health and environmental damage from chemicals of these different classes. The proportions of total pesticide use on cereals from these three groups is 5 per cent for insecticides, 25 per cent for fungicides and 43 per cent for herbicides. No EIQ information is available for growth regulators; this problem aside, a weighted average figure may be calculated for the New EIQ rating of the average pesticide:

$$\{(5*92) + (25*87) + (43*56)\}/(5+25+43) = 5043/73 = 69$$

If we take 69 as the mean new EIQ of pesticides applied to cereals, then it is clear that a pesticide with this EIQ should be associated with the average tax, here estimated as £12 per kilogramme. This suggests a tax per new EIQ point of 12/69 = £0.17 per point per kilogramme. That is, the differentiated tax as estimated here would be a tax of 17 pence per kilogramme of pesticides for each point on the new EIQ index which that pesticide had. Table 4.5 shows, in the column headed estimated tax, what the tax per kilogramme would be for the five insecticides considered.

We do not wish to stress the specific numbers arising from these calculations because of the many simplifications we have had to introduce in order to derive them. However, it is fairly clear what additional work would be required in order to produce more reliable results – as we feel would be necessary for any specific policy programme. In the meantime, however, we feel that these figures illustrate rather clearly two important principles. Firstly, they show that the level of external costs associated with pesticide use is potentially high, and that including these costs within the private costs of pesticides could have significant impacts both on the price of pesticides and consequently on the amount of pesticides used. Secondly, they stress the important point that the level of external costs associated with pesticide use is not constant for all pesticides. Some are more damaging than others. The figures in Table 4.5 show how important this could be for any taxation policy. By implication, it is clear how powerful an instrument differential taxation could be in ensuring that farmers make an efficient choice of which pesticides to use.

POLICY IMPLICATIONS: SUMMARY

Uniform pesticide taxes

The simple example above illustrates how a uniform pesticide tax rate could be calculated using survey data. The calculation would need to be corroborated by further research before strong policy conclusions could be derived. Nonetheless, the results suggest strongly that the marginal external costs of pesticide use could be significant.

Uniform pesticide taxes can be based on many criteria other than the measure of externalities: for example, meeting water quality targets, or raising revenues sufficient to implement clean up, education and training campaigns. Pesticide taxes of this nature are already in place in Norway, Sweden and Iowa, US. We argued in Chapter 3, however, that uniform pesticide taxes are unlikely to be the best policy for coping with pesticide externalities because of the differing marginal costs for different chemicals, and in different locations. So, while uniform taxes may be suitable for purely revenue raising purposes, they fail to encourage the efficient choice of chemicals, and fail to differenti-

Table 4.5 *Differentiated tax for insecticides used on wheat*

Insecticide	*Health*	*Ecology*	*New EIQ*	*Relative tax*	*Estimated tax*
Chlorpyrifos	27	105	95	£X*(2.2)	£16.15
Dimethoate	40	141	128	£X*(3.0)	£21.76
Esfenvalerate	6	137	120	£X*(2.8)	£20.40
Fonofos	26	83	75	£X*(1.7)	£12.75
Pirimicarb	23	46	43	£X	£7.31

Note 1: the slight difference in ratios between last two columns is due to rounding
Note 2: the relative tax column is new EIQ / new EIQ for pirimicarb

ate between those regions where the environment is highly sensitive and where problems are less likely.

Differential tax or subsidy policies using damage indices for pesticides

A better pesticide policy would distinguish between pesticides with different external impacts. One way to get more efficiency from pesticide policies is to make use of consistently derived ranking indices.

The tables above show how WTP estimates can be combined with scientific data to rank pesticides with different external impacts. However, if an average tax rate was estimated, the ranking index could be used to assign different taxes to different pesticides. Then the tax would have an impact not only on total levels of pesticide use, but also on the choice of pesticides. Farmers faced with a decision concerning which of two equally effective pesticides to use will clearly prefer the cheaper option, and differential tax rates could have a strong impact on this decision – especially given tax rates of the magnitude suggested by the survey. Were this idea to be adopted as the basis for a major pesticide tax policy, or for subsidization of less damaging chemicals, more research would be required to corroborate the initial results presented.

Differential taxation by location

We have not made any explicit calculations concerning differential taxation by location of pesticide use: this is for the simple reason that any such calculations would, by their very nature, be location-specific and thus not suitable for generalization. For the same reason, taxes which vary according to the location would be more costly to estimate and implement. However, it is clear how concerned policy-makers could adjust the new EIQ figures to reflect local conditions. For example, farms next to important bird breeding areas could face higher taxes for the ecological component of the new EIQ, while facing the same taxes as other farmers for the health component. This would require splitting the tax down more explicitly into the two components – which could be easily done.

Of course, there may be equity problems with taxation which vary by location in this way. These could be addressed, for example, by subsidies paid to farmers for adopting environmentally friendly practices in these areas, or by some kind of lump-sum rebate system which recycles some proportion of the pesticide-tax revenues to farmers in a way which does not vary with their use of pesticides, so that the incentives for reducing pesticide use remain the same.

Environmental damage indices as information for farmers

It may be possible to derive some environmental benefits without resorting to a taxation policy. Many farmers will feel motivated to protect the environment voluntarily to some degree. The ranking indices could be of use in providing better information to farmers about the environmental impacts of their activities, which would be of help in selecting pesticides and perhaps even in deciding whether or not to spray.

A further way of facilitating this decision process would be to combine the ranking indices with the economic injury levels (EILs), discussed briefly above as the current cornerstone of integrated pest management. Environmentally augmented EILs would give an explicit signal to farmers of when the economic value of crop protection justifies the private and external social costs of pesticide use. At present, EILs incorporate only private costs. Of course, compliance with EILs or environmental EILs is optional, and it is likely that taxation or some other policy would be required to reach an efficient outcome.

REFERENCES

Beggs S, Cardell S and Hausman J (1981) 'Assessing the potential demand for electric cars'. *Journal of Econometrics*, 16, pp 1–19.

Brouwer FM, Terluin IJ and Godeschalk FE (1994) *Pesticides in the EC*. Agricultural Economics Research Institute, The Hague, The Netherlands.

Central Statistical Office (1993) *Family Spending, A Report on the 1993 Family Expenditure Survey*. London, UK.

Department of the Environment (1996) *Pesticides in Water: Report of the Working Party on the Incidence of Pesticide in Water*. HMSO, London.

Foster V and Mourato S (1997) *Behavioural consistency, statistical specification and validity in the contingent ranking method: evidence from a survey on the impacts of pesticide use in the UK*. CSERGE Working Paper GEC, pp 97–109.

Garthwaite DG, Thomas MR and Hart M (1994) *Arable Farm Crops in Great Britain*. Pesticide Usage Survey Group, Ministry of Agriculture, Fisheries and Food, Herts, UK.

Higley LG and Wintersteen WK (1992) 'A novel approach to environmental risk assessment of pesticides as a basis for incorporating environmental costs into economic injury levels'. *American Entomologist* 38, pp 34–39.

Kovach J, Petzoldte C, Degni J and Pette J (1992) 'A method to measure environmental impact of pesticides'. *New York Food and Life Science Bulletin* 139, pp 1–8.

Lareau TJ and Rae DA (1985) 'Valuing willingness to pay for diesel odour reduction: an application of the contingent ranking technique'. *Southern Economic Journal* 55(3), pp 728–42.

5

Sustainable Business and the Pesticide Business: A Comparison

Dominik Koechlin and Anja Wittke

Introduction: Sustainability as a Corporate Strategy

Sustainable development has reached the corporate world. Many multinationals, the pesticide producers amongst them, have committed themselves to this key concept of the last decade. It is widely recognized that sustainability on a corporate level means balanced and durable economic, social and environmental performance. There is some dispute, however, on whether the two non-economic performance measures are important per se or whether they are subsets of the overall raison d'être of a company: *profit maximization*. The outcome of this discussion has important implications for the content of corporate sustainability.

If profit maximization is regarded as the prime company goal, then the scope for improvement in social and environmental performance is smaller than if economic, social and environmental performance are on the same level. In line with the macro-economic concept of sustainability, we could call the profit maximization priority the weak company sustainability, meaning that everything is done to improve social and environmental performance, as long as the profits of the company can still be maximized. For many companies, there is scope to exploit these win–win situations, for instance where increased pollution prevention has a positive impact on the bottom line. A *strong* sustainability concept would require that the economic, social and environmental goals of a company have equal priority. While it is clear that even in this case a company has to have high enough profits to sustain its development in other areas, strong corporate sustainability does imply that there are cases where the company decides to put environmental or social performance before profits.

In this chapter, we look first at concepts of sustainable industries, with lessons from other research-intensive industries. We focus on the pesticide industry and look for those key driving forces that are changing the traditional pesticide markets and threatening the sustainability of the industry. We then look at the strategies currently developed by companies to cope with that change, and finally assess those strategies for their contribution to 'weak' and 'strong' sustainability in terms of shareholder versus stakeholder value, with recommendations for companies which seek to put their visions for sustainable development into action.

Sustainable industries

A company's sustainability depends on the industrial framework in which it operates. Sustainable industries are industries where the benefits of the industry outweigh the costs. Crucial for an industry's sustained development is the perception of those benefits and costs by the company's customers along the *value chain* of the industry. Pressure for industry sustainability comes from different *stakeholders*, among them legislators, non-governmental organizations (NGOs), competitors, shareholders, company leaders, customers, suppliers and employees. A change in the way the stakeholders and players weigh the costs and benefits of the industry can result in a shift of the key factors necessary for success.

Sustainability can only be assessed by looking at the whole value chain of the industry. Are certain stages in the value chain under increasing stakeholder pressure? Does the industry have any influence on these pressures or is it affected by 'other people's problems'? A good example is the food processing industry which was heavily affected by the problems which the packaging industry created.

Before moving on, a look at some of the sustainability aspects of other industries should help prepare the ground for our analysis of the pesticide industry. The pharmaceutical industry is a good example of how a long tradition of innovation no longer seems to guarantee an ability to adapt to a rapidly changing environment (see Box 5.1). Innovation is not being induced by companies themselves but by major stakeholders, especially the health care providers, fuelled by the rising costs of the health care system. They will add to the sustainability of the industry only if pharmaceutical companies manage the transition from *product* to *value orientation* (see Table 5.1), in other words the transition from *pill producer* to a *health care provider* with a clear role within health care systems. Value orientation is regarded as being more sustainable because it enables the company to replace products with know-how wherever the market requires it.

Can the same be said of the pesticide industry? Can a shift from product to value orientation, and even to strong sustainability, avoid some of the painful lessons learned by the pharmaceutical industry and make a greater contribution to sustainable development?

THE PESTICIDE INDUSTRY AND ITS POWER IN THE AGRIFOOD SYSTEM

Since the 1950s the pesticide market has grown very rapidly and production capacities were installed all over the world. The 1996 world pesticide market was valued at $31.3 billion, up from $11.7 billion in 1980 (Agrow, 1997). Although pesticide producers have had good years in terms of sales volumes and profitability, the pressure on the pesticide industry is rapidly rising.

The pesticide industry is, by all standards, mature. Sales in the large markets have remained almost unchanged since the early 1980s. The main pesticide markets are still in North America, Western Europe and Japan, and the distribution of the market changed very little from 1985 to 1995. The growth in value between 1993 and 1998 is

BOX 5.1 LESSONS FROM OTHER INDUSTRIES

The pharmaceutical industry seems to be particularly relevant for learning more about the sustainability of industries and companies. Most large pharmaceutical chemical companies are involved both in the pharmaceutical and agrochemical industry, and the agrochemical industry is often affected by developments taking place in the pharmaceutical sector.

In recent years, pharmaceutical companies have gone through drastic changes. Health care reforms have increased the pressure on prices and margins. Generic producers are pushing into the market. Biotechnology promises new opportunities, but most traditional research and development (R&D) companies have had to buy new know-how, despite their ever increasing R&D budgets (estimated at $359 million per new product in 1990 by Kaitin, 1996). Recent acquisitions and moves of pharmaceutical producers to buy distributors and companies that produce over-the-counter drugs show that the players are trying to improve their position in an increasingly competitive environment.

This pressure is leading to drastic changes in the way pharmaceutical producers understand their business. In the US, where some of the most innovative health care organizations are located, a new approach is being used, where a health care organization does not just buy products or services to meet its customers' needs, it buys *performance*. An example would be that the organization responsible for the health care of a district would ask for a bid to stabilize the number of heart disease patients in that region by a certain percentage. Such a performance package could then include preventive measures such as anti-smoking campaigns and reduced fees for fitness centres, as well as efficient curative measures.

For pharmaceutical producers, the shift from a product and service orientation towards a performance orientation raises important questions. Are they capable of offering such a performance guarantee? If yes, will they make money out of it? What can they do if others provide the performance guarantee and use generic products to cut costs? The changes in the pharmaceutical industry have a lot to do with a shift in power within the health care system. Health care organizations and, depending on the system, hospitals and distributors are gaining power at the expense of the pharmaceutical producers.

Not many years ago one would have suggested that the sustainability of the pharmaceutical industry was secured thanks to a constant innovation and the very high perceived benefit of pharmaceuticals. Today the benefits are associated more with an efficient and effective health care system and less with specific pharmaceutical producers. Innovation is more difficult to achieve and many of the most promising innovations come from small companies or university researchers. Established pharmaceutical companies have had to pay high prices and large stock premiums to get access to innovative technologies and products.

estimated to be 4 per cent per year for the US and 4.6 per cent per year for Western Europe. These value gains will come from newer and more expensive products. But there are observers who are sceptical about this forecast. They predict a declining market because of the substitution of generics for existing products, because of the substitution of seed-based genetics for chemical technologies, and because farmers are

Table 5.1 *Contrasting product and value orientation*

Product orientation	*Value orientation*
• focus on products	• focus on customer value, utility
• large plants, high fixed costs	• products and services
• large investments in often marginal improvements	• investment in system
• high concentration in the industry	• new entrants and players
• low flexibility	• high flexibility
• sales and profits as key indicators	• shareholder value
• strategies focus on existing organization and structures	• strategies for creative networks (knowledge-based)

coming under increasing cost pressures and will have to reduce their input costs. What seems clear is that the period of continuing strong growth seems to be over, except for some parts of Asia and Latin America. But even in these growing markets, strong competition – increasingly from local producers – will keep prices down.

In an industry with strong growth, the 'pie' gets larger every year and therefore the fight for a nice piece of the pie is less intense than in a mature industry where market share can only be won at the expense of a competitor. In mature markets, concentration is likely to increase as cost structure and economies of scale become key success factors. Vertical integration increases, for instance through the acquisition of distributors or strategic alliances with suppliers, as does the concentration process through acquisition of competitors. Over the past decades the pesticide industry has become highly concentrated. Familiar and sometimes infamous names in the business have gone by the board – Shell, Fisons, Union Carbide, Schering, Hoechst, May & Baker, Rousell Uclaf, Ciba, Geigy, then Ciba-Geigy and Sandoz – to be replaced with unfamiliar names of the new conglomerates and joint ventures such as Novartis and AgrEvo. Today over three-quarters of the market is controlled by ten global pesticide producers, all based in Europe or the US (see Table 5.2). The newest and largest, Novartis, has combined pesticide sales of $4.5 billion, 50 per cent greater than its nearest competitor. The agriculture sector will witness more acquisitions and mergers because producers are actively seeking more balanced product portfolios, better geographical coverage and access to promising R&D pipelines. The agrochemical divisions are also under the influence of developments in the pharmaceutical industry, as shown by the acquisition of Cyanamid by American Home Products and the merger of Ciba-Geigy and Sandoz to form Novartis. Producers focus increasingly on their agricultural and health care businesses, through 'demerging' their chemicals divisions or forming agrochemical joint ventures. With the divestment of chemicals and focus on life-science businesses comes a tendency to claim the moral high ground of sustainability (*Chemical Week*, 1997).

Driving forces

Agricultural markets are strongly policy-driven. Over the last decades the agricultural policies in OECD countries concentrated on supporting farmers' incomes, fixing prices, subsidizing export sales, and restricting competing imports. The results were escalating costs of support programmes, international trade conflicts, environmental

Table 5.2 *Agrochemical sales of the leading pesticide companies*

Company	*Headquarters (agrochemicals)*	*Agriculture as per cent of company turnover*	*Agriculture R&D as per cent of sales (1994)*	*Pesticide sales in US$ millions (1996)*
1 Novartis[a]	Basle, Switzerland	28% (Healthcare 59%, Nutrition 13%)	9.3–11%	4,527
2 Monsanto[b]	St. Louis, USA	48% (Searle pharma 32%, Food ingredients 19%)	~5.7%	2,997
3 Zeneca	Fernhurst, UK	34% (pharma 45%, specialities 21%)	11.6%	2,630
4 AgrEvo[c]	Berlin, Germany	100%	12–13%	2,493
5 DuPont	Wilmington, USA	5% (fibres 17%, polymers 16%, petroleum 43%)	8.6%	2,472
6 Bayer	Leverkusen, Germany	10% (healthcare 26%, industry 18%, polymers 17%)	12%	2.360
7 Rhône–Poulenc	Lyon, France	12% (health 43%, chemicals 31%)	8.6%	2,210
8 Dow Agro Sciences[d]	Indianapolis, USA	100%	11%	2,000
9 American Cyanamid (AHP)[e]	Wayne, NJ, US	AHP: 12% (pharma 55%)	10%	1,989
10 BASF	Limburgerhof, Germany	6% (consumer products 22%, plastics/fibres 24%)	15%	1,541

Notes:
a In March 1996, Ciba-Geigy and Sandoz announced that they would merge as Novartis. Divestments of herbicides to BASF to avoid antitrust problems cost the new company about $320 million in sales in 1996. Speciality chemicals became an independent company in March 1997. Figures are for 1996, minus chemicals.
b In December 1996, Monsanto announced its intention to spin off its chemical business as a separate company. Figures are for 1996, minus chemicals.
c AgrEvo is the combined agrochemical interests of Hoechst-Roussel and Schering/NOR-AM.
d DowElanco was the joint venture between Dow (60%) and Eli Lilly (40%), consolidated by Dow. In May 1997, Dow announced that it would acquire Eli Lilly's stake for $900 million, making Dow the sole owner of the subsidiary, now named Dow AgroSciences. Dow AgroSciences accounts for 10% of Dow Chemical's turnover.
e American Home Products acquired American Cyanamid in November 1994. Figures presented are for AHP.
Source: PANNA from Agrow (1 March, 15 March, 29 March, 19 April, 1996; 18 April 1997). Agrow Top Ten, Company World Wide Web sites

market distortion and a farm sector maximizing output according to government policy and not market signals. Pesticide use has been correlated with the level of government subsidies. In the 1970s when subsidies were very high, the world market for pesticides grew by 7.8 per cent a year on average. During the 1980s, the market growth rate fell to a little over 3 per cent a year.

Government budget deficits and the World Trade Organization (WTO) and General Agreement on Tariffs and Trade (GATT) agreement make it impossible to support subsidies any longer. This trend is paralleled by the decline of farmers' importance to society and to the food system. Other interest groups that have a higher influence on policy-making will have a larger say in the forming of agricultural policy (Unnevehr, 1993). Agricultural policies were, and to a great extent still are, aimed at large industrialized farmers in order to secure their output maximization. Pesticide producers have focused on bringing products to the market that serve the needs of these farmers. Although the deregulatory trend in agriculture is driven by economics, it may at the same time also increase the sustainability of farming systems, making them less dependent on chemical inputs. Thus, *market distortions* which used to sustain the pesticide industry are diminishing.

Regulatory agencies play a significant role in pesticide-use decisions. They establish the parameters for the selection and application of pesticides and are increasingly involved with pest management issues. *Pesticide-use reduction programmes* are in effect in The Netherlands, Sweden, Denmark and Korea, though as the targets are set in volume reductions, these goals may be met in part through the natural market evolution to low-rate, high-efficacy products or, in the case of The Netherlands, the move away from soil disinfectants. Denmark and Sweden are moving to further restrict pesticide use through taxation (as high as 37 per cent in Denmark), bans or stringent reregistration programmes. Revenue from pesticide taxes in Denmark and Iowa, US is used to fund research and extension of more sustainable farming systems.

Free trade is also disrupting the complex system of *price differentiation*, enabling parallel imports of pesticides, a trend which will have a strong impact on future cash flows in the industry. Even in Europe prices for products with the same active ingredient can vary by a factor of two, and prices in Europe can again be ten times higher than prices in the developing world. Many *patents* will expire in the near future and the slowing rate of innovation cannot stop the trend of growing market share of off-patent products. Generic producers are disrupting long-established markets by offering the same products at lower prices, a fact that will rapidly gain importance in an increasingly cost-conscious agrifood industry. In 1993, 14 per cent of the West European pesticide market was accounted for by generic products from generic producers. AgrEvo expect this figure to rise to 25 per cent by 2000, and the share of off-patent proprietary products to rise from 33 per cent to 50 per cent. One forecast for the US predicts sales of generics to account for 67 per cent of the market there by 2005 (*Farm Chemicals*, 1996). The pesticide producers can lose twice over from generic manufactures, through the loss of export markets in the developing world and through the loss of market share in the markets of the industrialized world. China now produces over 200,000 tonnes of pesticide active ingredients, and in India more than 125 companies produce over 60 technical grade pesticides totalling 86,000 tonnes (*Agrow*, 1996).

The environmental pressure on the pesticide producers remains very strong. There is no doubt that the current debate on produce quality and pesticide use is partly the result of the efforts of non-governmental environmental groups and consumer organizations. The power of NGOs in the debate on pesticide-use policy has increased. The Pesticide Action Network played a substantial role in achieving the prior informed consent (PIC) provisions. The position of many NGOs active in agricultural policy can be summarized by the following quote: 'As voters, citizens and taxpayers, we cannot allow this widespread destruction (food contamination, loss of livelihood) of our environment to continue for the next 30 years. We must oblige politicians to develop a new ecological direction of agriculture' (Greenpeace, 1992). The NGOs' position is supported by increasing public scepticism about whether the current registration system is able to protect us from health hazards and whether the environmental impact of pesticides is taken seriously enough.

Although new products using much less active ingredients per acre are on the market, regulators in all major markets are increasing their demands, making new product developments ever more costly. Research and development costs have risen enormously in the last years. In 1975 R&D costs per product were around 25 million ECU ($32 million) and in 1992 they grew to 125 million ECU ($158 million). The costs for studies of toxic and environmental impact rose from 4 million to 50 million ECU ($63 million), and industry leaders are questioning whether current levels of R&D expenditure can be sustained. Costs of R&D are so high that once a product has been developed, a worldwide market is required to ensure profitability.

Increasing pressure from regulators, although not always sustainable in terms of its balance between economic efficiency, efficacy and environmental requirements, has put the environmental performance of pesticide products at the top of the agenda for manufacturers. The complicated regulatory system for pesticides imposes large costs on the pesticide producers, but because of the high margins in the industry they are still bearable and increasingly serve as a barrier to entry by new competitors. In nearly all OECD countries, there has been a noticeable trend towards combining agricultural and environmental policies, and rural development is also entering the agricultural policy debate. The use of traditional regulatory approaches, such as pollution emission standards or quality targets for the environment and product standards for pesticides, is still very common. Unfortunately, economic instruments that discourage harmful environmental impacts and promote beneficial environmental practices or effects have rarely been implemented. With the orientation of many government agencies towards the whole agrifood chain, the links between food safety, health and environmental measures are of growing importance in agricultural policies.

Pesticide regulations have not managed to raise the confidence of the end consumers in the safety of pesticides (see Figure 5.1), even though an increasing percentage of the registration costs for a new pesticide goes into environmental studies and the registration procedures become increasingly more complicated. A survey conducted by the industry association ECPA revealed that, on average, 80 per cent of people questioned in 11 European countries were unaware of the fact that pesticides have to go through a complex registration process (*Agrow*, 1994).

The reputation of pesticide producers in many countries is still not good. The industry is seen as a major polluter and pesticide residues in food are prominent on the environmental agenda. The pesticide industry has failed to correct this image. Certain industry arguments – for instance, the agrochemical industry 'feeds the world' – have been counterproductive because it is quite clear that intensive farming in the industrialized world can never be a solution to the food problems in the developing world. In the developing world, pesticide producers are perceived as forcing Western farm technology and inputs into an environment for which these products are not suited. The responsibility for the correct application of pesticides in these countries increasingly lies with pesticide manufacturers, raising their costs for information and the training of farmers within product stewardship programmes. Another driving force is the question

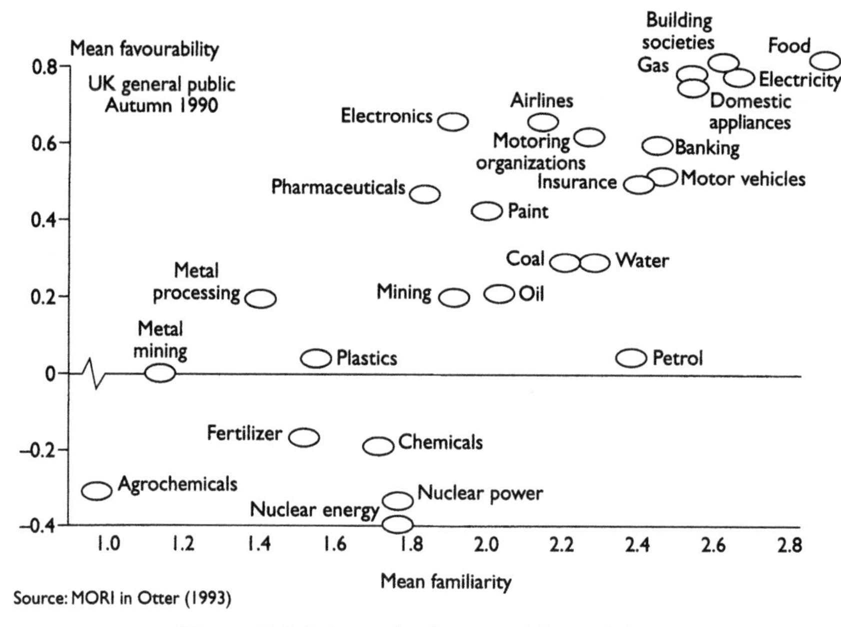

Source: MORI in Otter (1993)

Figure 5.1 *Industry familiarity and favourability*

of potential liabilities of the pesticide industry in a post-Bhopal world.

It seems questionable whether the new products the pesticide industry has to offer are optimized along the value chain. The changes taking place in the pesticide industry can only be understood if one looks at the entire agrifood chain and the main players influencing its trends. If the power shifts within the value chain, then agrochemical products have to take the perceptions of the other players into account. If stakeholders of the industry are changing their needs and their perceptions, we must ask whether the industry is able to adapt to these changes and create shareholder value while improving its environmental and social performance. As we saw with the brief analysis of the pharmaceutical industry, changes in stakeholder influence on the industry are crucial.

Players and power in the agrifood chain

Stakeholders in the agrifood system today

The agrifood chain includes upstream suppliers of raw materials, the farm sector, distributors, food processors and food retailers. Agricultural policy and other sectors of public policy, such as environmental policy and food safety, form the policy framework.

As Figure 5.2 shows, there has traditionally been a strong link between pesticide producers, distributors and farmers. This link was well suited to the push strategies of pesticide producers, which focused on farmers and enabled the pesticide producers to earn high margins. The downstream sector, such as food processors, retailers and consumers, has not had a large influence on agricultural inputs. The marketing efforts and the relationship management of pesticide producers, therefore, did not extend beyond farmers along the value chain.

Independent private wholesale and distribution companies and cooperatives handle most of the distribution of pesticides. In most countries, they are the sole link between

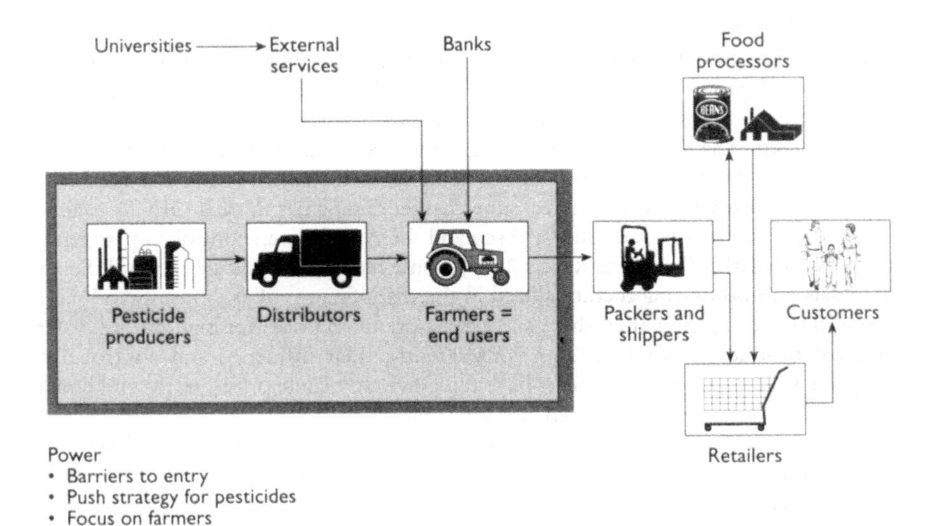

Figure 5.2 *The agrifood system – past*

manufacturers and farmers. The bulk of their income often comes from buying and selling the farmer's products. Pesticides often account for only a small part of their sales, but for a much higher part of their profits. In the EU, the share by cooperatives of pesticide markets varies from about 15 per cent in the UK to about 50 to 60 per cent in Germany, France and Italy (Dutch Agricultural Economics Research Institute, 1994). The strong link between manufacturers and wholesalers has been the basis for the push strategies of pesticide producers has enabled them to gain good market penetration.

In maturing markets, the distributors gain increasing importance as gatekeepers to the market and a good link with distributors is one of the industry's key success factors. Competition increases dealer power to demand service and guarantee product performance, which puts pressure on industry costs. If the distributors' needs are not satisfied, they will not push a product through to the farmers. With more complex products, the importance of retailers rises, because the sale of the product is dependent on the services that can be provided. For the service-intensive integrated pest management (IPM) techniques, distributors and manufacturers will have to share the task of educating farmers, which will make the ties between pesticide producers and distributors tighter and the potential for conflicts concerning the share of value higher. When power shifts in the agrifood chain and marketing strategies fluctuate from push to pull, the close link with distributors is crucial for getting access to market signals.

The farm sector has been declining over the past decades. In the US, the farm sector's share of total economic value of food declined from 41 to 9 per cent. In contrast, the input sector's share rose from 15 to 24 per cent, while that of the marketing sector increased from 44 to 67 per cent (Schaller, 1993). The decline in the farm sector has gone hand in hand with a concentration process; in the US the largest 15 per cent of farms (about 265,000 farms) now account for at least 80 per cent of all gross cash-farm commodity sales. There are basically two strategic alternatives for farmers: they can rely on further cost cutting and become cost leaders in their markets or they can focus on special niches, where the pressure on prices is less intense. Both alternatives will be heavily influenced by the requirements of food processors and retailers, which reflect consumer perceptions. In this process, farmers will lose part of their decision-making power to the players downstream. Increasing downstream concentration by food processors and retailers is the result of business strategies aimed at capturing economies of scale, increasing domestic market control and competing in world markets. The top five manufacturing firms in a broad food industry sector generally account for more than 60 per cent of total domestic market share. The food retail sector is dominated by a handful of supermarket companies in each OECD country. For instance, in Britain, Sainsbury's, Tesco and the Argyll group control 40 per cent of the national market (Lang and Raven, 1994). Some retailers now closely control the raw materials and processing methods used in the manufacture of in-house brand-name foods in order to gain market share and extract premium prices for high-quality, differentiated products (OECD 1994). Consequently, their influence and control over suppliers is growing. Today retailers have the power to dictate prices, terms and conditions under which crops are grown and delivered.

Retailers do not have the same agricultural expertise as farmers. Their main concern is not with cultivation and the economic survival of farmers, but with supplying competitive products. Competitive farm products are products that meet consumers' demands in terms of price, quality, convenience and food safety. The latter is mainly a question of perception. As we will see, a rising number of consumers, depending on market and product, perceive pesticides as something they would not like to have in their food. Food processors and retailers get more and more involved in the 'making' of the products they buy from farms. Cooperatives and packers play an increasing role in pesticide decisions. For example, Sun World and Di Mare Brothers, two of the largest packers in California, have established their own brands. They provide growers with detailed instructions and are developing private extension services to advise farmers. OECD-sponsored studies suggest that food companies are moving into plant and animal breeding in order to 'programme' the desired characteristics of raw products (OECD, 1994).

There is a growing market for food that is produced under special production standards. In most countries retailers have tried two different strategies for capturing this market. The more successful strategy has been to introduce their own label for food grown under an IPM scheme. Organic food has been much less successful in most supermarkets, perhaps because consumers demanding organically grown food do not look for it in their local supermarket. An exception is Coop, the second largest Swiss retailer, which has a very successful brand for 'naturally grown' food and now faces the challenge of satisfying a growing demand. This involvement is only partly related to potential market benefits which guaranteed pesticide-free food could provide. With the processing and marketing of farm products rapidly gaining in importance, retailers, and especially food processors, are increasingly reluctant to risk having their image destroyed by pesticide residues found in their produce. Raw material costs can be as low as 5 per cent of the end costs of a branded product, and therefore processors and retailers are willing to pay a premium to eliminate the residue risk. This opens new possibilities for IPM or organic farming, because the success of branded products is no longer solely dependent on the willingness of end consumers to pay a price premium, but also on the need of food processors and retailers to reduce perceived risk. For example, baby food producers are very strict regarding the specifications of the products they use; the German manufacturer Hipp only uses raw materials from organic farms to eliminate any risk for its brands. Hipp has invested in education of its contract farmers, and its own advisors help farmers in the cultivation process. For quality control, an impressive infrastructure of high-technology and company laboratories was set up (Hipp, 1996). Because an increasing number of food processors and retailers are buying their produce through direct contracts, their pressure on farm practices and pesticide use will only increase.

Of course, the same influence will also affect farmers' decisions to plant genetically engineered varieties. Food processors and retailers will not fly in the face of consumer concern, irrespective of the claims by the biotechnology and seed companies that there are no human health risks with the new varieties. A contradictory impact of the concentration process on pesticide use is the increasing availability of out-of-season food and more uniform products. A large percentage of consumers, though health-conscious, reject farm products that do not adhere to the highest cosmetic standards, which makes a pesticide use-reduction strategy difficult for the supermarkets to implement on a wide scale. There are few niches for farmers to supply supermarkets with crops that are more environmentally suited and require less pesticides.

Changing consumer preferences during the 1990s have taken the form of increased demand for lower-priced products, but longer-term trends are towards higher quality, more diverse, convenient, healthy and environmentally friendly products (OECD, 1994). Some of these characteristics conflict with lower prices. Retailers will pass the problem on to the farmers and demand all the characteristics consumers are looking for at lower prices. It is interesting to note that the increased value from organics goes to the downstream sector and the farmer gets only a small portion of added value while bearing higher production costs.

Consumer surveys and retailers' experience have shown that there is a large gap between the environmental awareness of consumers and the premiums they actually pay for environmentally safe products. Consumers are unlikely to accept a rise in food prices in real terms. Over the last decades, expenditure on food has fallen constantly as a proportion of disposable income; in the US, this drop is from about one third of income spent for food in the 1960s to 11.4 per cent in 1994, 40 per cent of which is consumed away from home. The farmers' share of the food dollar is declining; in Germany in the 1950s, 75 per cent of a German mark spent on food went to the farmer, compared to less than 20 per cent today. All consumer surveys reveal that pesticide use is considered a major issue for food safety. In many countries there is a strong scepticism that the problem is dealt with adequately. In the US, 59 per cent of the respondents to a major national survey on meal planners strongly to mildly disagreed that 'the current law adequately protects me from eating foods with dangerous amounts of pesticide residues in them' and a majority of respondents (64 per cent) strongly to mildly agreed that 'pesticides should not be used on crops grown for food because the risks are greater than the benefits'. Over 50 per cent of respondents preferred to buy organically grown fresh fruits and vegetables and the same percentage said that they would pay more for produce that was certified as pesticide residue-free (Lynch and Jordan-Lin, 1994). It seems clear that these findings have consequences for the sustainability of a pesticide producer's strategy.

Interactions in the agrifood system

The concentration process along the whole agrifood chain continues. The concentration of retailers and their power to fix prices and conditions under which crops have to be grown and delivered, and their large demand for food, offers many farmers no alternative but to adhere to their standards, except for some small niche markets. The changes of power in the agrifood system are summarized in Figure 5.3.

These developments have a contradictory impact on pesticide use. On the one hand, food, quality, standardization and cosmetic appearance require higher pesticide

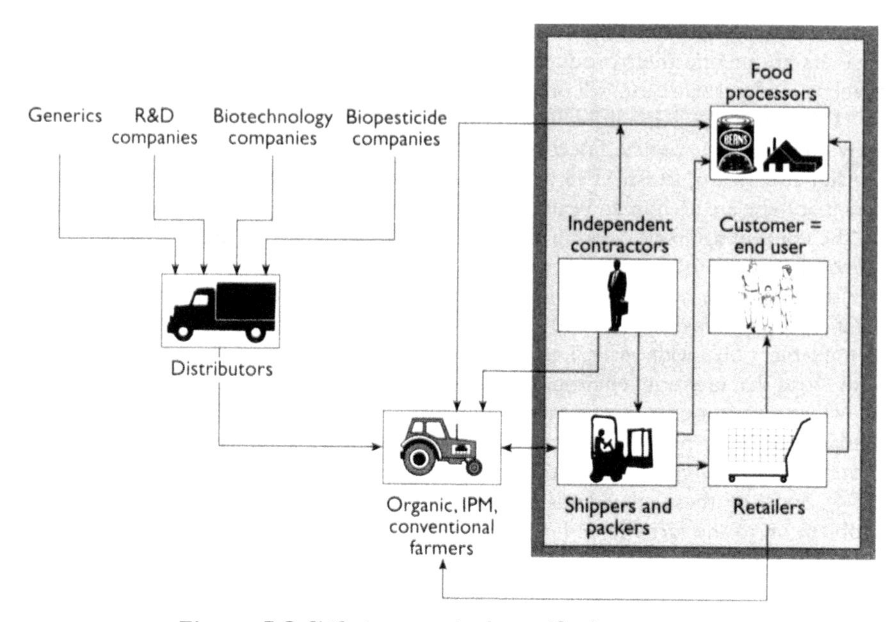

Figure 5.3 *Shifts in power in the agrifood system*

input. On the other hand, consumer perception and buying behaviour reveal growing concern about pesticide use. The efforts of the players in the downstream sector (the creation of labels, stronger standards for pesticide use, or the monitoring of foods for pesticide residues) are indicators of the government's declining credibility to regulate risk in food. It is not clear whether the current trend – government regulations replaced with standards set by retailers – will change consumers' perceptions about modern agriculture. One possibility is that new players such as independent organizations or contractors will establish a new network amongst the players and take responsibility for safe food. From the perspective of farmers and retailers, a contract could minimize market risk and increase their credibility with consumers.

Our analysis of the driving forces shows that market conditions in the agrifood chain have changed dramatically. The development of agriculture in the industrialized world has resulted in a shrinkage of the role farmers play in that system. The power in the agrifood system has shifted away from the pesticide producers and farmers, what one could call the input side of the agrifood equation, to distributors, shippers and retailers, or what might be called the output side of the agrifood system (Unnevehr, 1993). The pesticide market has consequently shifted from a seller's to a buyer's market.

> *What you're seeing is not just a consolidation of seed companies, it's really a consolidation of the entire food chain. Companies like ours, who want to continue to be in the food and feed production business, are all trying to secure our spot along that chain.*
>
> Robert Fraley, President of Monsanto's Ceregen Division
> (*Farm Journal*, October 1996, p19)

A summary of the key political, economic, social and technological influences on the pesticide industry is presented in Table 5.3. It is interesting to note that – contrary to the views of most environmentalists – it is hard *economic* influences on the industry

Table 5.3 *PEST analysis of the pesticide industry – major influences and drivers of change*

Political/legal	*Economic*
GATT/free trade Ecotaxes/use-reduction targets Declining political support for agriculture	Commoditization/market maturation Competition for market share Patent expiration, generics Shifts in power in agrifood system ➩ retailer influence Farming ➩ low-margin business Environmental liabilities (chemicals and biotechnology)
Socio-cultural	*Technological*
Public resistance to chemicals and biotechnology Consumer preferences: price, quality, convenience, health, environment	R&D costs Market saturation Pesticide resistance Biotechnology (herbicide-resistance, pest-resistance)

which are the greatest threat to its sustainability. The question is whether the current strategies being employed by pesticides producers can respond to these challenges in ways which improve their own sustainability and the sustainability of the agricultural sector which they serve.

CURRENT STRATEGIES OF THE PESTICIDE INDUSTRY

What strategies are being employed by producers in the pesticide industry to meet the challenges of industry maturation, public disquiet and shifting power in the agrifood system? In this section, we look at research, product, market, and 'communication' strategies currently employed by the pesticide industry.

R&D strategies

The agrochemical industry spends nearly $3 billion per year on research and development. As the number of companies that are able to afford the rising research costs diminish, so are many companies stressing the need to achieve a 'critical mass' of around $150 million in research expenses, amounting to 10 per cent of turnover, as a basis for a successful R&D programme. To reach that point, strategic alliances between competitors are quite common. For Schering, the joint venture with Hoechst (AgrEvo) was a way to profit from another company's R&D efforts.

Policy efforts such as the US-EPA's reduced-risk programme, where new products are registered more quickly if they are 'safer' than existing alternatives, will be an opportunity for proactive companies. There are great variations in the producer's product portfolios, but most of them anticipate a future for environmentally compatible products and the development of new technologies to improve IPM techniques.

Biotechnology and seeds

For pesticide companies, investment into biotechnology and seeds is driven by a double urgency – to exploit the new market of protection via the seed, and to compensate for the predicted impacts of the new biotechnologies on existing core chemical pesticide markets, which could be reduced by $820 million annually in the US alone by 2005 (*Farm Chemicals*, 1996).

The first generation of crop protection biotechnologies has, after huge investments in research, brought two types of crops to the market. The first is herbicide-tolerant varieties of maize, oilseed rape (canola), soybeans and cotton. The most exciting of these varieties, such as Monsanto's 'Roundup Ready' soybeans and AgrEvo's 'Liberty Link' system, can free the farmer from the intricacies of using selective herbicides, with their narrow crop tolerances and application windows, and risk of crop damage. Instead, farmers can pass over a mixture of weeds and crop with a broad-spectrum, 'burn-down' herbicide such as glyphosate or glufosinate, and every plant will be killed except the crop, which has engineered resistance to the herbicide. The developer of the technology recoups the R&D costs either through a 'technology fee' included in the seed cost (for instance, Monsanto and Novartis) or indirectly through sales of the herbicide (for example, AgrEvo).

The potential for accepting the herbicide-tolerant crop technology concept among farmers – who with this simple system can now spread their management over even more acres – and the threat to established chemicals, is dramatic. Herbicide-tolerant maize hybrids are expected to account for around 15 per cent of US maize acreage by 2001. In soybeans, weed control with Roundup may cost as little as one fourth of a normal soybean herbicide regime for the season. Some forecasts for Roundup Ready soybeans in the US predict ten million acres (from a national total of 64 million acres) to be planted in 1997, eventually leading to a 30 per cent decline in existing US soybean herbicide markets. The irony of this cannibalism of the chemical pesticide markets with biotechnology is that the segments most affected are those on which chemical screening has been focused: herbicides in maize and soybeans and insecticides in cotton.

The other group of the first wave of crop protection biotechnology are the 'Bt' insect-resistant varieties. These are engineered to express an insecticidal protein from the bacterium *Bacillus thuringiensis* (Bt), so that insects which are susceptible to Bt are killed when they feed on plant tissues. Pesticide companies actively developing this technology include Monsanto and Novartis, starting with cotton, maize and potatoes. The move into Bt varieties allows a company to take a short-cut to a big chunk of the major insecticide submarkets in cotton and potatoes. By contrast, the Bt maize market in the US does not replace conventional chemical treatments, as only 2 to 3 per cent of the 60 million acres infested with corn borers is currently treated with conventional insecticides. Cash grain farmers in the US have been very wary of insecticides because of their higher mammalian toxicity and environmental impact, and it is interesting to note that when Iowa farmers were asked what they perceived as the greatest advantage of Bt maize, the majority of growers (41 per cent) said less insecticide in the environment, 21 per cent said less insecticide exposure to farm workers, and 20 per cent said potentially higher yields (*Farm Industry News*, 1996). Growers can expect to pay a premium of $5 to $10 per acre or $15 to $30 per bag for Bt maize. We will soon be witnessing the 'stacking' of genes conferring tolerance to a number of herbicides and insect control with Bt within single varieties.

Most companies with pesticide divisions bought into the seed business in the 1970s and 1980s, but the race to profit from crop biotechnology has led to a rash of large investments to guarantee access to agricultural biotechnology and to seed markets. In one two-year period, Monsanto spent nearly $2 billion on strategic seed and biotechnology investments. Development costs of biotechnology are less per product than for chemical pesticides thanks to the easier registration hurdles. But barriers to smaller companies still exist, in terms of access to patents, and the movement of pesticide companies into biotechnology seems likely to hasten the concentration process:

> *A handful of large companies claim most of the patents on plant biotechnology. According to Mike Sund, director of corporate communications for Mycogen, the number of basic technology players is probably down to no more than a dozen major corporate players. Carlton Eibl, president and chief operating officer of Mycogen, thinks the number of major seed players will be even smaller by the end of the century, down to four or five.*
>
> *Progressive Farmer* (1996)

Of course, potential exists for a large independent seed company to introduce pest resistant varieties, and even give away the pest resistance for free without charging a premium over the nonresistant seeds, just to maintain market share.

The focus of agricultural biotechnology will shift from extension of the pesticide market (herbicide-tolerance) through substitution technologies (pest-resistant varieties), to agronomic characteristics, such as drought, or salt or cold tolerance, and an outside chance of nitrogen-fixation in cereals, as well as biopharmaceutical products (such as human vaccines or pharmaceuticals from plants) and industrial products such as lubricants. The boundaries between crop protection and seed production are set to blur.

AgrEvo expects the world market for genetically engineered plants to be worth $6 billion by 2005, equivalent to 20 per cent of value of current global pesticide business. The European agricultural biotechnology market is forecast to expand by 23.6 per cent per annum from 1995–2002 (to $3 billion, from $890 million in 1995 and $490 million in 1992). But note that agricultural biotechnology is only 3 per cent of the EU biotechnology market (compared to 68 per cent for healthcare and 27 per cent food processing), which will expand to only 3.7 per cent by 2002. The forecasts for the market potential of these engineered crops vary dramatically from low to very high growth rates, depending on predictions of market acceptance by the end consumers.

The launching of agricultural biotechnology has been controversial. Consumers are asking what benefit they get from having genetically engineered products in their food supply. Surveys in the US, Australia, Canada and Europe reveal that at least half of the population are very concerned about the risks from genetically engineered foods. The launch and poor market performance of Calgene's Flavr Savr tomato has shown that there is considerable public scepticism towards biotechnology, especially in Europe. It is astonishing that pesticide producers, having lived with public objections to chemical technology for so long, seem once more to be carried away by the possibilities the new technologies offer, rather than responding to consumers' perceptions.

Biological control

The industry's love affair with genetic modification of crops has rather eclipsed an anticipated move into biological control products, especially microbial pesticides (bacteria, viruses or fungi which are applied against target pest insects, diseases and weeds), behaviour-modifying chemicals (usually synthetic versions of insects' sex pheromones, which are used to trap or confuse pests), and natural enemies (parasites or predators of pests which are usually reared and mass-released). The appeal of biologically based pest-control products is considerable, especially the much less stringent (and consequently cheaper) registration requirements, faster time to market, and positive public relations value (OTA, 1995, p 157). The development cost for biopesticides is around $5 million and the time to market is about three years compared to over $100 million in development costs and eight to 12 years for conventional chemical pesticides. For many scientists and observers of the pesticide industry, any shift to technologies for sustainable agriculture *must* involve considerable investments in biocontrol and biopesticides.

Nevertheless, very few biological products have made it into the product range of pesticide manufacturers, and even fewer have flourished. Companies have been confronted with technical and commercial features which make these products difficult to distribute, sell and use alongside traditional chemical products (OTA, 1995, p157), these features include:

- Short shelf lives (especially natural enemies);
- Narrow spectrum of activity;
- High requirement for technical support;
- Variable quality;
- High production costs.

These factors limit biologicals to high-value niche markets where growers have high levels of technical support. Exceptions are microbial pesticide formulations based on Bt, which can be stored and applied like a chemical. The world market for bacterial biopesticides was expected to increase from $68 million in 1992 to $275 million in 2002 (*Pesticide Outlook*, 1994). But the interest of pesticide companies in biological control appears to have been a transient phenomenon, nipped in the bud by a R&D focus on new reduced-toxicity chemical pesticides and biotechnology.

Restructuring and Cost-Cutting Strategies

Costs, especially R&D costs, are rising and margins are slipping as more markets become cost driven. Cost-cutting is a major issue in every company's strategy and it will gain in importance since cost leadership is regarded as the best way to protect off-patent products. Most companies have started to focus their R&D portfolios and to reorganize the whole company. A few such as Shell have taken steps to downsize or divest their agricultural business. A characteristic of all restructuring projects is the aim for flatter hierarchies. With massive reengineering efforts, the pesticide producers try to adapt their structures and processes to the new environment.

Ciba believes that cost-cutting is vital to maintain profit growth in its crop protection division, and the merger with Sandoz will cut costs via shared promotional activities and reduction in personnel. The CEOs are predicting savings of SFr 1800 million ($1435 million) from the merger within three years of the merger approval, including SFr 420 million ($335 million) in the agribusiness sector. Monsanto invests several million dollars every year on making cost improvements and is committed to being the lowest cost producer worldwide (*Agrow*, 1995c). Du Pont's cost-reduction strategy has resulted in total savings of $3000 million since 1990 (Agrow, 1995b). In 1994, Zeneca sold its Garden Care line because this business was no longer core to their main activities in the international agricultural market. Cost-cutting strategies are necessary to secure profitability in the future but they run the risk of creating a *risk-averse corporate culture* which jeopardizes the company's ability to exploit new business opportunities.

Product strategies

The market for conventional pesticides is dominated by established, effective and cheap products, many of which have been around for over 25 years. There are, however, still markets with high value, such as the fruit and vegetable market, which has grown over the last ten years. Fruits as well as vegetables generate a high value added per hectare. Competition will be fierce, but as long as there is market growth, profitability will remain high.

Different trends can be observed for maintaining or increasing market share:

- *Product stewardship* can differentiate even the oldest products from generic competition. Zeneca's leading herbicide product has been off patent since the late 1970s and is on the Pesticide Action Network's 'Dirty Dozen' list of most dangerous pesticides, but its sales volume has doubled in this time. This is due to the investments in manufacturing and formulation and innovative marketing (*Agrow*, 1995a). Monsanto forecasts sales volume of their successful product Roundup – which brings in 15 per cent of Monsanto's income – to guarantee its continued growth. It has been reducing the product's price and investing in safer formulations in order to make the market less attractive for its competitors, especially the generic producers. Monsanto still has 95 per cent of the world glyphosate market.
- *Cost leadership* is the traditional way for protecting off-patent products. Sandoz, on the other hand, has focused on their highly strategic products, reducing the product portfolio from 20 to nine, and concentrating its resources on higher-quality standards. The success of setting new quality standards is mostly a question of price and market acceptance. Moreover, if the market and the influential regulatory agencies are interested in safer products, the pioneer has a good opportunity to make large profits while forcing the followers to meet the costs of reaching the new standard.
- The *exchange or licensing of products* for completing product portfolios is increasing. Cyanamid and Sandoz have agreed to introduce a combination of herbicides in the US based on each other's proprietary products for use on maize and soybeans. Rohm & Haas acquired Monsanto's pyridine herbicides that complete Rohm & Haas's speciality crop business and no longer fits Monsanto's core row-crop focus.
- There is an increasing focus on *niche markets*, such as seed treatment. Zeneca, Bayer and Ciba-Geigy are the most successful seed treatment companies among the ten largest pesticide producers. The earning potential of these niche markets is estimated to be around 10 per cent of crop-protection sales by the end of the decade, so they will remain niche markets but with a high profitability potential. Being active in different niche markets reduces the pesticide producers' risks of investment failures but also increases the risk of not having the critical mass for exploiting market potential.
- *Creating own-label generics or acquiring a generics producer*, a strategy quite common in the pharmaceutical industry, can be a way of holding on to the markets for off-patent products and not losing them to generic producers. This strategy has already been implemented by Rhône Poulenc, through the acquisition of Interphyto, and by AgrEvo in 1994, through the purchase of the German generic specialist Stefes. The reason for this acquisition was that 50 per cent of AgrEvo's pesticide sales come from products that no longer are under patent protection and are exposed to increasing competition from generic firms.

Generic and off-patent strategies for traditional research-intensive companies cause major rethinking of corporate raison d'être and culture.

> *Most of those employed in crop protection have always been excited and motivated by product innovation and it is an integral part of the industry's culture. This is changing to some extent as the industry tries to get more mileage out of old products, although many executives will find it hard to adapt to a market that is moving more towards a commodity-trading mentality.*
>
> (*Crop Protection Monthly*, 1992)

Our review of current industry strategies has revealed a distinct focus on the *economic* third of sustainable development, in other words a distinctly 'weak' corporate definition of sustainability. Companies still have a predominantly product-orientation, despite the trend towards substituting crop biotechnology for chemical inputs. The focus on technology has distracted companies from addressing shifts in power in the agrifood system. The manner in which biotechnology development is following the old technology-driven chemical strategy shows that most companies have yet to develop a value-orientation and to address issues of social sustainability.

IMPLEMENTING A SUSTAINABLE STRATEGY IN THE PESTICIDE INDUSTRY

We wrote in the introduction that corporate sustainability has to encompass *economic*, *environmental* and *social* aspects. A company's management has to be able to assess, to plan and to manage the company's performance in all three dimensions. Looking at corporate sustainability, therefore, has a lot to do with tools and indicators that enable the company to institute its values.

A sustainable strategy for a pesticide producer should include the following elements (see Figure 5.4):

- knowledge management;
- strategic planning and focus;
- environmental management;
- value management – shareholders and stakeholders.

Knowledge management

The term knowledge management describes what is, in our view, a prerequisite of sustainability on a corporate level. Knowledge will increasingly replace capital as the most valuable asset of many companies, especially pesticide producers. Although it often only shows partly on the balance sheet, for instance when an acquisition allows for goodwill to be activated or when brands are capitalized, knowledge is often worth more to a company than its physical assets. With active development and interchange of knowledge within the organization, and with its 'network', the company can create the constant change that is needed for sustainability. Knowledge management in this sense reduces the risk associated with operating in a rapidly changing environment. The challenge of knowledge management lies in transforming the company's knowledge into sustainable value creation.

Two types of knowledge are commonly observed:

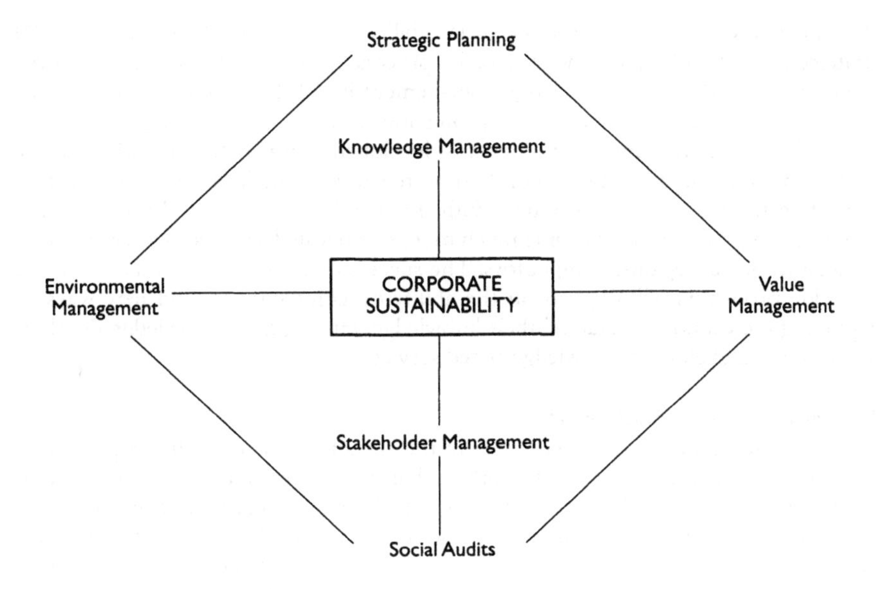

Figure 5.4: *Tools for corporate sustainability*

- *Migratory* knowledge moves across company boundaries and can be acquired easily.
- *Embedded* knowledge defines how a particular company conducted its business, its corporate culture, etc.

For corporate sustainability it is crucial to actively manage both types of knowledge. The company has to have a clear view of the migratory knowledge available, has to assess whether knowledge must be developed in house or acquired, and has to constantly search for new combinations of knowledge and 'package' them into products and services. Embedded knowledge has to be managed in the sense that new knowledge must diffuse through the company to be used by all the company's businesses; if a new marketing approach was developed for one business unit, it could be adapted for others as well. For corporate sustainability, embedded knowledge in the sense of corporate culture and 'the way things are done' is vital, because values such as sustainable development have to evolve over time in order to be a guideline for the company's operations.

The problem with implementing knowledge management is that management systems such as financial controlling are still based on a different organizational view of the company. They guarantee that control can be exercised through the hierarchy but do not foster knowledge development and distribution. Investments in knowledge such as R&D and training are treated as costs, not as investments for the future which could depreciate over their lifetime. But knowledge management requires that power is not exerted hierarchically; rather, it is delegated to the lowest possible level. This systematic installment of entrepreneurship within large companies makes a common vision, such as a sustainable strategy, a hard task for top management to implement. It is important for corporate sustainability to develop management and information systems that encourage self-control and the systematic development and exchange of knowledge.

For pesticide producers, knowledge management is especially crucial. Many of the elements that could add to corporate sustainability are knowledge-based, such as IPM strategies, biotechnology and the registration process. For pesticide producers, however, the move towards active knowledge management is a difficult one. In the past the processes of knowledge creation and protection were well known. Large sums were invested in R&D, and once a product was developed, it was patented and therefore protected. Once the patent expired, there were several generic strategies for further exploiting the value of the product. With the new knowledge-based products and services, these mechanisms become much more complicated, as most services are very difficult to protect against competitors. The success of a sustainable strategy for the pesticide producers will depend on the ability of companies to find new ways of exploiting the economic value of their knowledge, ranging from knowledge development to the marketing of knowledge-based services.

Diversification to service-based products

Diversification plays an important role in the sustainability of pesticide producers because of the uncertainty of their markets. This is not diversification in the sense of entering completely new businesses. Rather, it is diversification as a way of participating in new technologies and markets without carrying all the risks associated with developing the expertise and potential business in house. Depending on the tradition of pesticide producers, diversification can mean a major shift in their strategy.

The trend in the last ten years has been that pesticides have become more sophisticated, meaning that the amount of active ingredient has declined and the knowledge needed for their application has increased accordingly. The service content of the product has risen, and if alternative farming practices such as IPM, where pesticides are reduced thanks to increased know-how, are applied, the importance of the services provided will further increase. Would it therefore be a solution for the pesticide producers to seek their income from *providing services*? Probably not. Farmers are used to getting the services of the producers for free and will be reluctant to pay for them in the future (see Chapter 8). In addition, there is the issue of protecting services. Once a service is provided, it can easily be copied and local consultants could be in a much better position to provide the services in the future. A service-intensive IPM strategy is not easy to implement on a large scale. What can be done on a carefully surveyed trial farm is not necessarily transferable into a mass product. And the training necessary for IPM can only be provided if the distributors are locked into the system through higher margins or other incentives.

This raises the important issue of scale. A sustainable strategy has to deal with the conflicts arising between standardized, global products and differentiated local needs (den Hond et al, 1997). A sustainable strategy will require pesticide companies to face the trade-offs between being large-scale, cost-efficient producers and more specialized, focused producers. The limitations of service-based strategies do not dictate that a sustainable strategy for pesticide producers has to focus on selling physical products. For certain applications of pesticides, the notion of *insuring* the crop is important. If the crop is used as a collateral for financing the farm (as is often the case in the US), or if the subsidies are based on product output, it is vital for farmers to be insured against crop loss. Do pesticides have to provide this insurance? There are other ways of insuring a crop, depending on the nature and the probability of the damage that is threatening the crop. The simplest way would be to offer insurance coverage, just as insurance is sold against hail damage.

Our look at the pharmaceutical industry highlighted a new combination of products and services called *performance contracting*. A performance contract goes much further than the services accompanying a product or the insurance-related services discussed above. When a farmer buys a performance contract, he is guaranteed a certain performance for a crop. The contract specifies the amount and the quality of the crops guaranteed and the fee the farmer has to pay. The contractor specifies in a detailed plan what the farmer has to do and stays in close contact with the farmer during the growing period. It is in the interest of the contractor to advise the farmer in the most cost-efficient way in order to maximize his margin. For the farmer, a performance contract adds value because the fee for the contract is lower than the farmer's costs without the contract. Contractors can offer this lower fee because they have the know-how and the resources to manage the crop in the most cost-efficient way.

From a sustainability point of view, performance contracting can have several drawbacks. The farmer loses a large amount of decision-making freedom to the contractor. The contractor can be induced to make the farmer use cheap and environmentally less-sophisticated pesticides in order to keep costs down. On the other hand, the contractor is a highly specialized company and can use the most up-to-date expertise and technology to limit the environmental effects of farming. The contractor reduces the farmer's risk for the guaranteed crop, which enables the farmer to take higher risks with other crops while having a guaranteed source of income. Contracting requires a new combination of knowledge. It includes elements of insurance, crop management, global sourcing for cost-efficient inputs, financing and so on. A company offering contracting services will have to enter alliances with specialized companies in these areas.

Because of their background, pesticide producers would be in a good position to offer performance contracts. The question is whether they are perceived as being independent and flexible enough to apply not just their own products, but to choose the most effective input-combination available. Performance contracting is an attractive concept. To make it profitable, there are still a lot of problems to be solved. But it seems clear that the pesticide producers have to carefully think about the possibilities performance contracting could offer to their business of selling sustainable agriculture.

Strategic planning and focus

Sustainability in a corporate sense means having a clear view of the company's vision, planning its strategy and maintaining a constant readiness to adapt to new developments. Because of the new driving forces and fundamental changes taking place in the industry, the importance of having a clear strategy increases. But strategic planning has been out of favour in recent years because it is not able to prepare companies for the rapid changes they experience in the business environments (Mintzberg, 1994). There are very few markets left for which the prerequisites of conventional strategic planning – stable markets, familiar competitors, long product life-cycles and slow technological change – occur. But the relegation of strategic planning is very dangerous for sustainability because strategic planning could be the only activity where the company's top managers can think long term. To sustain economic performance as a pesticide producer, it will no longer be sufficient to have a good product pipeline and to be a successful marketer of pesticides.

Alliances in the agrifood chain

A consequence of the change in power is that the bargaining position of pesticide producers will weaken, and with it their share of the value created in the agrifood system. Increasing competition in the agrifood system means that the players in the system have to compete harder for their share of the value. The change in power will require new alliances with those players in the agrifood industry who have the power. The strategies of the pesticide producers will have to focus not on adding value for the farmer, but on maximizing value for the powerful players in a specific market, which is a quite different task from pushing new products to the farmers. Pesticide producers will have to start talking to food processors and retailers. A crucial element of the marketing of pesticides will be the management of the whole agrifood chain. It will be especially important for those crops where either the product the consumer buys is more or less unchanged as it comes from the farm, such as fruit and vegetables, or for those products where the processing and marketing make up the bulk of value added, such as food with strong branding – for instance, breakfast cereal.

Pesticide producers could offer:

- know-how for the specification of contracts with farmers;
- analytical capacity for quality control;
- guaranteed residue-free food where appropriate;
- guaranteed product stewardship for the pesticides used;
- shared consumer information on farm production and health issues;
- shared 'perception management' by increasing the transparency of the industry.

The question is whether food retailers and distributors can be allies in a sustainable strategy. The best chance of winning them as partners for sustainability lies with those pesticide producers that give them high incentives in terms of higher margins. Because it is hard to add value in the agrifood system, this would suggest that giving away higher margins means a smaller shareholder value for the pesticide producers. This could be a first indication of the trade-offs of a corporate sustainability strategy.

Strategic alliances with competitors

Pesticide producers have to note the changes that scale and time have on their strategy. High R&D costs required worldwide launch of products. The time horizon for R&D was ten years from the discovery of a new molecule to a new product on the market. If markets are stable, the high investment and long time horizon are not worrisome. If markets are changing rapidly, then committing R&D money can be a risky business, because the market might have very different needs by the time the products are ready. One way of becoming less dependent on time and scale is to form a system of strategic alliances. Strategic alliances mean sharing potential risks, but also potential profits with partners. This can lower the profits of the pesticide producers, but will increase their chances of survival.

Strategy for developing countries

The strategies of pesticide producers are focused on the industrialized world. From a sustainability point of view, the developing world should gain importance. The problem is that most farmers in the developing world lack the resources to buy external inputs. Companies have viewed the small farmer markets of the developing world as issue-rich and profit-poor. But selling to these markets with a hands-off attitude is not sustainable, especially with older, more toxic products. There seems to be a clear lack of ideas on how to serve the needs of the developing world as a pesticide producer. The current practice of primarily serving the triad of the industrialized countries and using basically the same approaches in the developing world wherever possible not only risks missing business opportunities, but also makes the companies vulnerable to stakeholder pressure. A sustainable strategy would have to address the specific issues of the Third World and develop solutions matching the specific needs of developing countries. In certain parts of the Third World there is misuse and overuse of pesticides. Productivity is low, despite relatively high input costs. Trials have shown that IPM strategies can result in higher yields for the farmer than conventional pesticide use. The farmer has a strong incentive to adopt IPM and the effects for the environment are positive. One current approach by Novartis – a dedicated international farmer support programme which takes a long-term perspective to sustaining markets through skills building – shows considerable promise.

When to lead

Companies that take the lead in an industry can force competitors to keep pace and can alter existing standards of environmental performance to improve the industry's sustainability. The question of whether or not to take the lead with sustainable performance in an industry is a difficult one. It depends on the company's assessment of stakeholder pressure for sustainability and on potential market opportunities created by sustainable products and services. The issue of timing is especially tricky when margins are still high. Large companies are risk-averse and when the advantages of making the first move cannot be adequately quantified, the status quo prevails.

There is also the question of benefits. If a company has a strategy of sustainable development, the chances are high that a large amount of the benefits will go to what economists call free-riders. This is an important issue for companies because there are many who could benefit from a corporate sustainability strategy, including the company's competitors. If a sustainability strategy makes competitors better off in the short term, for instance when market share is sacrificed in segments which are seen as unsustainable, the competitors could use the short-term advantages to win customers that are hard to retrieve.

Organizational redesign

A sustainable strategy requires an organizational redesign. Corporate structures have to give the managers in charge the flexibility to offer their customers the product or service combination required by the market. This can only be done if the structure of the organization delegates power to the lowest level possible and at the same time allows the strategic vision for the company to move in the direction of a sustainability strategy. Adequate incentive schemes and controls have to be installed to enable the corporate vision to be implemented without hindering the initiatives of line managers. Corporate reengineering programmes that have been launched by most of the large players will have to find ways of making the processes of the organization more efficient and effective without losing strategic orientation.

Environmental management

Environmental management is the general term for management systems that deal with a company's environmental performance. Three categories of environmental management tools can be identified:

- *Eco-controlling* is a system for continuous improvement of the environmental performance of a company's activities. The methodology used is very similar to management accounting. Eco-controlling is often used to manage the environmental impact of production at a specific site, but can also be applied to a whole company. Eco-controlling normally deals with processes, not with products, and the system boundaries are often identical with the factory gates.

- *Life-cycle analysis* is the methodology for analysing and comparing the environmental impact of different products, product variations and packaging materials over the entire life-cycle of the product. By definition, life-cycle analysis goes far beyond a company's range of influence.
- *Environmental audits* traditionally were conducted to make sure that a company was complying with complex environmental legislation. Audits were done periodically and not continuously and their focus was more on the legally relevant environmental risk of the production processes than on continuous reduction of pollution.

Today the lines between these three environmental management tools are becoming less clear. The term audit is also used for continuous management tools and efforts are being taken to include product aspects into eco-controlling. There is also a strong tendency to coordinate and harmonize environmental management systems through international standards (such as ISO, EU, BSA etc), though the effect on the efficiency and effectiveness of environmental management systems remains unclear.

Environmental management systems play an important part in a sustainability strategy. But their use to date has unfortunately been mostly by production engineers to improve the production process. Only when a variety of environmental management tools become part of managers' tool kits throughout a company will the environmental requirements of a sustainable strategy be met.

Value management – shareholders

Companies were formed for economic reasons. Their aim was to create value for their owners: the shareholders. Once they grew to a certain size and expanded worldwide, it became generally accepted that they had additional responsibilities. The sustainability discussion on a corporate level is about what a company's responsibilities are and how it can cope with them.

Value management deals with the economic foundations of a company. It is the main tool for quantifying strategic alternatives and the trade-offs that are part of every sustainable strategy. The shareholder value concept reveals several drawbacks of traditional accounting (Rappaport, 1986):

- Accounting figures do not include risk.
- The time value of money is neglected.
- The influence of accounting standards varies.
- Accounting figures describe the history of a company and not its future.

There are several important elements for assessing the shareholder value of a company or a business unit, the most important being the *future free cash flows*. Free-cash flow is defined as the cash flow that is at the disposal of the company's fund providers, whether equity or liabilities. The free-cash flow consists of the company's cash flow minus the investment in working and fixed capital. Company-specific, so-called 'value drivers', such as sales growth, incremental investment in working capital and tax rate, can be identified and their influence on the value of the company quantified.

The second crucial element of the shareholder value concept is the weighted average cost of capital, which defines the cost a company has to pay for its capital, equity and debt. The weighted average cost of capital depends on the risk of the company which is assessed by the stock markets. What makes the cost of capital so important is the fact that investors discount future cash flows using the cost of capital as the discount rate. The rationale is that a dollar today is worth more than a dollar tomorrow, because if an investor has the dollar today, he can start earning money with it. Managers use the cost of capital as a hurdle rate in the sense that an investment only adds to the company's value if its return is higher than the cost of capital. The higher the cost of capital, the higher the discount rate and the less importance is given to cash flows in the distant future.

Shareholder value analysis is all about looking at the future free cash flows the company can earn and discounting them into a value which is expressed in today's dollars. Value management is therefore focused on future cash flows, the associated risks and the time horizons. These issues influence the corporate sustainability debate. High costs of capital mean shorter time horizons. But does value management add to a company's sustainability? We first look at the other elements of corporate sustainability before taking a closer look at the possible contributions of value management to corporate sustainability.

Value management – stakeholders

A large company clearly has social implications for the people, the communities and other organizations that come into contact with it. There is no consensus as to what the social responsibility of a company should be or what would comprise the appropriate management tools. Some companies focus the management of their social performance on employee satisfaction and community involvement and donations. Others have very strong social principles, such as lifelong employment, without ever communicating their social management code. Recently there has been a new attempt to develop the management systems for the social aspects of sustainability, known as *stakeholder management.* A company can be regarded as a function of its stakeholders' demands. Stakeholders provide the company with resources and in exchange expect resources from the company. Many stakeholders can express their demands either by market power, for instance as customers, or through the political system of the country where the company is located. Large companies have an increasing number of stakeholders that either do not have enough economic clout to express market power or are too small to influence the political framework. Nevertheless, these stakeholders have to be taken seriously as their actions can erode a company's activities if their demands are not somehow integrated in the company's strategy.

Managing value within a system such as the agrifood chain is much more dependent on the perception of the end consumers than on selling products to one element in the chain. Implicitly, it also means parting from the focus on products. The benefits of crop protection products are often not well perceived. Management of perception in a strategic sense means much more than PR campaigns. Managing perception means active response to stakeholder needs. It requires a constant dialogue with the stakeholders and includes the communication of the industries' 'problem areas', but also the willingness to listen to demands for change. Working closely with their 'customers' customers' is the only way for pesticide producers to build credibility. Although pesticides have a bad reputation with the greater public and with important stakeholders in major markets, the industry is risking, with biotechnology, a repeat of its earlier mistakes. The new opportunities make many companies oblivious to the fact that it could be the *perception* of the new technology that is vital to its economic success. A greater sensitivity to demands of downstream stakeholders in the food chain could lead to a reevaluation of the costs and benefits of biotechnology. The crucial point for stakeholder management is that stakeholders have a variety of demands. Theory suggests that companies should weigh the stakeholders' demands according to the power of the stakeholders. In practice, it is very difficult for the management to know what stakeholders' demands are and to find a balance amongst the different demands.

The key controversy? Shareholders versus stakeholders

Our analysis has shown that the shifting power of stakeholders along the agrifood chain is rapidly changing the business climate for producers of pesticides and agricultural biotechnology. But how closely should a company respond to those stakeholders, especially if that response (such as diversification away from chemicals) threatens shareholder value? This is a key controversy. A company committing itself to a corporate sustainability strategy will have to face criticism that it is not maximizing shareholder value and that it is using managerial and financial resources to meet *stakeholder* demands. There are many reasons why the company should focus on *shareholders*:

- Shareholder value is the best indicator of a company's economic well-being. If the company is economically healthy, then the stakeholders receive the maximum benefit.
- Integrating stakeholders' demands into a company's strategy and operations erodes the competitiveness of the company. Competitors that focus on shareholder value will be more successful.
- If the company does not maximize shareholder value, its ability to attract capital in the future will decline.
- Stakeholder demands often cannot be quantified and their effects on the bottom line are too complex to manage.

From a value management point of view, this criticism can be related to the value drivers affected. Not maximizing shareholder value could mean that a company has higher costs of capital, which has an immediate effect on the company's value through the discount rate. It could also mean that the company, because of stakeholder demands, chooses to stay in a business that has less growth potential than other potential businesses and that free cash flows are lower than they could be. Often a company will find that stakeholder demands, such as pressure from environmental groups or new legislation, will require additional investment, which again lowers free cash flows and affects the value of the company. Of course, there are also cases where stakeholder demands are synergistic and increase shareholder value, but we will focus on the area of conflicting interests because this is where a corporate sustainability strategy is tested.

In order to gain a better understanding of the shareholder–stakeholder trade-off, it is useful to understand the characteristics that underlie their different positions. They can be summarized in terms of:

- time horizon;
- mobility; and
- measurability.

Investors are often accused of having short time horizons, while most stakeholders demanding corporate sustainability have a long time horizon. Finance theory, which provides the basis for value management, suggests that the time horizon of shareholders is also long, because a large part of the shareholder value consists of free cash flows generated in the distant future. The problem of different time horizons lies, therefore, not with fundamentally different objectives, but with what is called 'information asymmetry' between managers and shareholders. Because shareholders do not have the necessary information, they have difficulties assessing the company's long-term prospects and therefore concentrate on the short term. Managers wanting to please their investors often increase short-term performance and information instead of focusing on the long-term prospects and on building trust with investors.

Mobility plays a key role in stakeholder demands. Shareholders can vote with their feet. If they are not happy with a company's strategy, they sell their shares. Most stakeholders are more closely tied to the company and are therefore dependent on the sustainability of the company's strategy. *Measurability* is key for the stakeholder–shareholder controversy, because managers think in economic terms, which means costs and earnings. Stakeholder needs are often not coherently articulated and are hard to quantify in terms of costs and benefits. These key differences can be less critical depending on the so-called 'corporate governance'. Corporate governance describes an increasing trend where shareholders, who are, after all, the owners of the company, have increasing influence in the company's management. They articulate their concerns and launch proxy fights at annual meetings if the management does not react to their pressure. Involved shareholders are more likely to think long term. The effect on corporate sustainability can be positive or negative, depending on the values the shareholders represent. A growing percentage of shares on all major financial markets is held by institutional investors. Many of them represent pension funds, which by definition should focus on sustainable strategies. The reason why they often do not, at least in the US, is that they are constantly underfunded, so they have to maximize short-term earnings to fund their pay-outs.

How can managers find the balance between shareholder value and conflicting stakeholder demands? A first step should be to make the trade-offs transparent. The benefits of stakeholder management have to outweigh the costs of a lower shareholder value. Costs of lower shareholder value can be financing (cost of capital, access to capital) or operational (lower sales and margins, higher costs and investments). For instance, an active IPM strategy might deliberately shrink markets and reduce sales but guarantee long-term presence of products in a more sustainable market. The argument that benefits of stakeholder management should be reflected in the future free cash flows of the company leaves out the fact that shareholder value management does not capture all the economic benefits of a strategy. It does not account for the economic value of a strategy's *flexibility*. New models are being tested for a better analysis of strategic decisions, but what is clear is that there are economic values for the company that are not adequately reflected by shareholder value analysis, and the flexibility provided by good stakeholder management could be one of them.

Once these issues are made transparent, managers can actively communicate them to shareholders, especially pension fund holders, who should have the same interest in long-term sustainability. The financial and environmental reports should be replaced by corporate reports targeted at stakeholders. The shareholders' report should include the environmental aspects that are important for the company's stakeholders and should highlight the costs and benefits of a sustainable strategy. A report for the stakeholders with a special interest in the environmental performance of the company should include the company's economic possibilities for achieving sustainability.

Another key controversy of stakeholder and shareholder management is the role of *diversification*. Modern finance theory suggests that shareholders are not interested in corporate diversification because they can diversify their portfolio much more cheaply than can the company. If they wanted to invest in a 'sustainable agriculture' company, they would look to a specialist company rather than to Monsanto or DuPont. Many stakeholders, on the other hand, are interested in corporate diversification because they are much more attached to the company and often cannot diversify to reduce their risk exposure. The company's employees are good examples.

The models developed by modern finance theory are based on the assumption that financial markets are 'perfect' markets. In many countries this is not the case. For example, because of high transaction costs (Coase, 1937) there can be good reasons, even from a financial point of view, to diversify the company's business portfolio, in which case shareholder and stakeholder needs are much closer than finance theory would suggest.

Conclusions

We have not set out to address the impact of pesticides on agricultural sustainability. Our analysis has instead focused on the sustainability of the pesticide industry itself, considering the massive shifts in demands from stakeholders along the agrifood chain. Our review of the pharmaceutical industry showed how such shifts can tip the balance of power in ways which are detrimental to technology-driven, product-oriented companies, and can cause painful and costly readjustments. We have described similar shifts underway in the agrifood system, and noted that the current strategies of pesticide companies – concentration, cost-cutting, new chemistry and biotechnology – fail to adequately address these trends.

Our point is that a pesticide company which pays close attention to the management of *knowledge*, *strategic focus*, and *environmental impact* – of production and products – will greatly increase its flexibility and ability to respond to these trends in ways which improve the company's value to both shareholders and stakeholders. This can be a means to escape the creeping commodification of the chemical pesticide industry. Moving into knowledge-based pest management services, for example, could be both profitable and responsive to food retailers' demands for greater control over the production and use of inputs in food production.

But the prognosis for a large-scale shift of pesticide manufacturers to 'strong' sustainability – and thereby a closer compliance with corporate visions of sustainable development – is limited by weak market signals for technologies and services for low-input agriculture. These signals in turn reflect consumer demand for cosmetic food quality and year-long availability of food items, and the constraints on innovation in farming itself. The contradictory signals which the agrifood system sends to pesticide producers clearly need deeper investigation to unravel the complexities of stakeholder demands. This is the purpose of the next two chapters.

References

(1994) *Agrow: World Crop Protection News*, 203, p6.

(1995a) *Agrow: World Crop Protection News*, 222, p7.

(1995b) *Agrow: World Crop Protection News*, 224, p3.

(1995c) *Agrow: World Crop Protection News*, 228, p15.

(1996) *Agrow: World Crop Protection News*, 263, p18.

(1997) *Chemical Week*, 21 May, p 3.

Coase RH (1937) 'The nature of the firm'. *Economica*, 4, pp368–405.

(1992) *Crop Protection Monthly*, November, p1.

Dutch Agricultural Economics Research Institute (1994) *Pesticides in the EC.* Research Report OV 121, April 1994. The Hague, The Netherlands.

(1996) *Farm Chemicals*, March, p78.

Greenpeace (1992) *Green fields – Grey future: EC Agricultural Policy at the Crossroads.* Greenpeace International, Amsterdam, The Netherlands.

Hipp (1996) *Environment Report 1995*, Hipp KG, Koordination Umweltschutz, Pfaffenhofen, Germany.

Hopkinson P, Shayler M and Welford R (1994) 'Environmental policy, legislation and business strategy: the case of the transport sector' in R Welford *Cases in Environmental Management and Business Strategy*, pp 136–151. Pitman Publishing, London.

Johnson J and Scholes K (1993) *Exploring corporate strategy*, 3rd ed. Prentice-Hall, Hemel Hempstead, UK.

Kaitin KI (1996) 'Pharmaceutical innovation in a changing environment', in P Davis (ed) *Contested Ground: public purpose and private interest in the regulation of prescription drugs*, pp 109–123. Oxford University Press, New York.

Lang T and Raven H (1994) 'From market to hypermarket'. *The Ecologist*, 24 July/August 1994, pp 124–129.

Lynch S, Jordan-Lin CT (1994) 'Charting the costs of food safety: meal planners express their concerns'. *Food Review*, May–Aug 1994, pp 14–22

Mintzberg H (1994) *The rise and fall of strategic planning.* Prentice-Hall, New York

OECD (1994) *Agriculture Policy, Market and Trade, Outlook and Monitoring 1994.* OECD, Paris.

OTA (1995) *Biologically Based Technologies for Pest Control.* Report OTA-ENV–636. US Congress, Office of Technology Assessment, Washington, DC.

Otter J (1992) 'Some aspects of environmental management within a chemical corporation' in Koechlin D and Müller K (eds) *Green business opportunities: the profit potential*, pp 81–98. *Financial Times*/Pitman, London.

(1994) *Pesticide Outlook*, February 1994, pp 4–5.

(1996) *Progressive Farmer*, June 1996.

Rappaport A (1986) *Creating Shareholder Value.* The Free Press, New York.

Schaller N (1993) 'Farm policies and the sustainability of agriculture: rethinking the connections'. *Policy Studies Program Report No 1*, Henry A Wallace Institute for Alternative Agriculture, Maryland, US.

Unnevehr LJ (1993) 'Suburban consumers and exurban farmers: the changing political economy of food policy'. *American Journal for Agricultural Economics*, 75, pp 1140–44.

6

UNRAVELLING THE STAKEHOLDER DIALOGUE OF PEST MANAGEMENT

Mick Mayhew and Sam Alessi

Talking is like playing on the harp; there is as much in laying the hands on the strings to stop their vibration as in twanging them to bring out their music.

Holmes, *Autocrat of Breakfast Table,* Chapter 1

INTRODUCTION

There exists a debate among people of differing walks of life about pesticide use and the sustainability of agriculture. In this chapter, we take the first step towards unravelling the stakeholder dialogue that surrounds the pesticide issue. Our motive is to identify and scrutinize innovations within the rigours of sustainability. Gaining an understanding of the underlying social patterns among stakeholders can lead to sustainable innovations that increasingly satisfy the public dialogue that surrounds the pesticide debate.

Consider the following questions: if you wanted to develop a sustainable political or business strategy for pest management, how would you do it? Maybe you would hire experts to find a solution. Perhaps you would talk with consumers and others who are concerned with the issue. Maybe you would attempt to involve both groups and revisit the underlying assumptions of the original goal. Would you argue that a new strategy really is not needed? If you decided to interact with concerned people, which of the stakeholders would you talk to? Farmers? Corporate types? Entomologists? What would happen if you added an environmentalist into the mix? Or would you avoid it because of the potential conflict and complications it would bring? What about including a politician? An economist? How about an urban-dwelling food consumer? What happens when you put these people in a room for a conversation? What would arise from and encourage a conversation? New ideas? Maybe.

If you envision this as a difficult situation, you are not alone. Although such a discussion would be extremely difficult, it must not be ignored if the development of sustainable strategies is desired. In fact, the new and holistic view of sustainable development requires input from a broader sphere of influence and perspective. If the pesticide industry wants to innovate within the principles of sustainable development, it must know how to: encourage greater stakeholder participation; guide the stakeholders through an information-gathering process; and manage the abundance of acquired information into usable business strategies. The search for new business opportunities is complicated for those in business and industry, and sustainable ideals are often perceived as a threat to the existing infrastructure. This perception often leads to fear or denial that any problem or need for change exists. Meanwhile, an attempt to climb the walls of difference is further exacerbated by the extreme complexity of the very *humanness* of individual and social artefacts.

The purpose of this chapter is threefold. Firstly, the reader is introduced to a way of thinking about human complexity that occurs in multistakeholder focus groups. We believe that a theoretically based and rigorous method of inquiry is necessary to maximize the outcome of focus groups dealing with social complexity such as pest management. Secondly, the reader is given a sample of results gleaned from our method of stakeholder analysis. Thirdly, based on our stakeholder analysis, implications of this approach for the pesticide industry will be offered. We hope that our approach will encourage sustainable technological innovations that integrate economic, environmental and socio-cultural issues – replete with all their complexities.

Stakeholder Focus Groups

Focus groups are a frequently used means to determine what people want in a product or service. Focus groups are generally organized and conducted by professional facilitators familiar with human interaction and group methods. There are many styles of focus groups and while most of them are based on similar principles, they differ in their capacity to solve problems of increasing stakeholder complexity (Flood and Jackson, 1991). Problems usually incurred when using traditional focus-group methods involve loss of information, problems with group dynamics, lack of a formal description of the human interaction, and an inadequate means of managing the large amount of information generated. In response, we feel compelled to discuss some of the human principles that are important in focus groups, for example, problem complexity and stakeholder infighting.

Any innovation, design or development of a policy, product or service requires knowledge of the *substance* that makes up the product. To determine this substance, some method or *process* must be enacted by *people*. This substance–process–people (what, how, who) triad is fundamental to human work and becomes increasingly difficult to manage as complexity increases. Badaracco and Ellsworth (1991) noted a conflict between emphasizing the 'what' and the 'how' of making decisions. Focus groups are particular events where this triad is clearly evident and where improper application of substance, process or people can lead to problems. For example, a common phenomenon today is the overemphasis of process with an associated underemphasis on people and scrutiny of substance.

Stakeholder knowledge

One way to increase the emphasis on people is to ensure that the participating stakeholder group is representative. This is done by simply asking yourself and others, 'Who has a vested interest in the area of concern?' The stakeholders are those people who are in the problem context; the people who 'language' about what they call a problem; and the people who comprise the social system (Anderson and Goolishian, 1988). Adherence to this focus group principle can greatly increase the completeness of the information that emerges from the analysis.

Involving a variety of stakeholders is important since each individual that participates in the group discussion brings a knowledge that is wrapped up in his or her personal experience, bias and self-interest. This knowledge can be divided into *embedded knowledge* and *migratory knowledge*. Embedded knowledge 'resides primarily in specialized relationships among individuals and groups and in the particular norms, attitudes, information flows, and ways of making decisions that shape their dealings with each other' (Badaracco, 1991). Embedded knowledge is generally difficult to obtain, but when it is disclosed by a stakeholder, it can provide previously unexpected opportunities for new innovations. A window is provided through which others can peer and see the world within the sociocultural context of that particular stakeholder.

A second kind of knowledge called *migratory knowledge* is more familiar to people and is easier to manage since it can be 'clearly and fully articulated–it resides in tidy, mobile packages like books and formulas, in machines, and in the minds of individuals' (Badaracco, 1991). Most focus groups deal primarily with migratory information that lacks controversy among stakeholders. The migratory knowledge is the main source of descriptive information, and when exchanged, provides cross-fertilization *between* stakeholders of ideas by which synergistic innovations can emerge. The combined knowledge of stakeholders is necessary for new ideas to emerge, but as the group interacts the need for change also becomes evident. Change is a developmental process that occurs over time. In focus groups, two human developmental processes occur simultaneously: *individual cognitive development* and *development of the stakeholder mind.*

Necessity of individual stakeholder development

Individual cognitive development requires changing one's belief structure – something that stakeholders frequently require of others but not of themselves! People become accustomed to what is familiar and prefer not to venture very far from it. For example, we enter into a relationship with someone who has a similar interest or experience. We may also enjoy this person if they have unique qualities that spark our interest. However, if we do not see a similarity, or their qualities are extremely foreign to ours, we feel uncomfortable, find a reason to dismiss them, and foreclose on the relationship. This is similar to how we respond to new information (Piaget, 1952, 1955).

Change requires the cognitive integration of new information. When we are exposed to new information, for example through print, conversation, or physical experience, we look for pegs within our mental maps on which to hang this new information. Our mental maps are a type of cognitive scheme that is structured by experiences over the course of our lives. When exposed to new information, we look for familiar pegs that can help us interpret this information. If the information is foreign to our usual way of thinking, no familiar 'peg' exists. Unless an individual maintains a stance of curiosity (Alessi and Mayhew, 1995), the meaning of the new information will likely be displaced by the familiar. For example, upon completion of one person's thought, another person may respond to a particular word or phrase, yet miss the entire meaning. Because the responder was unfamiliar with the meaning, he or she may search for something within the other person's statement upon which to respond. This tends to redirect the conversation into an area where the responder has some knowledge that is drawn from his or her own experience. If this redirection goes unnoticed, it will change the trajectory of the entire conversation. This is not a problem in casual conversation, nor is it typically intended, but it does prevent the responder from learning something new, which – in turn – prevents cognitive development and change. Deflection of new information in individuals is problematic to stakeholder dialogue in focus groups because an integral part of group discussion is that each individual adapts new information into his or her mental framework. It must be noted that maintaining an openness to new ideas does not require one to embrace them.

In their studies of wisdom, Orwoll and Perlmutter (1990) refer to the ability of an individual to 'move beyond individualistic concerns to more collective or universal issues', or *self-transcendence*. Self-transcendence requires the moral maturity (Kohlberg, 1973) of an individual to go beyond an egocentric focus, enabling the person to gain a deeper understanding of philosophical and epistemological issues (Orwoll et al, 1990). The individual may not agree with what is being introduced, but has at least respectfully taken the time to understand what the other person is saying. Once new information is accommodated into an individual's cognitive map, he can enhance the discussion with an informed and relevant response that reduces the conflict and adds robustness to the dialogue. The focus group facilitator should establish an atmosphere that invites self-transcendence among stakeholders.

Development of the stakeholder 'mind'

In addition to individual cognitive development, the second developmental process occurring simultaneously in a stakeholder group is the development of the stakeholder mind. The term 'mind' is not used merely to represent the brain in someone's head; rather, it is the result of the interaction of a human system. It is the meta-mind consisting of all the stakeholders acting as a group. Bateson (1979) describes mind as the real thoughts, beliefs and relationships that exist in the real world among the involved stakeholders. Physicist Paul Davies (1992) in his discussion of the deeper level of explanation, simply defines mind as the 'conscious awareness of the world'. Once understood, the stakeholder mind is fertile ground for discoveries that can lead to sustainable innovation.

Language is the 'data' by which mind, knowledge, difference and change become evident and meaningful. The words we use provide symbolic ways to transmit meaning within our socio-cultural networks (Maturana, 1978; Maturana and Varela, 1987; Bateson, 1979, 1971; Bateson, Jackson, Haley and Weakland, 1956). Meaning and understanding are socially and intersubjectively constructed within social systems (Anderson and Goolishian, 1988). While this dialogical crossing of perspectives occurs within the focus group, Braten (1987) cautions that conversation can collapse into a monoperspective. This is only a problem if the monoperspective prematurely emerges before the concerns of the stakeholders are fully understood.

Meaning is not static, rather its motion is fluid. As people interact, their cognitive schemes are challenged. As they seek to reduce the consequent discomfort, they look for new meanings that may require integrating the challenging statement into their scheme. As a result, change occurs within the individual who then responds in a different way than earlier. This sets off a domino effect of new challenges to stakeholder beliefs within the language system. Although the changes are minute and usually unobservable, the gradual adjustment of meaning that the language system makes can have a significant effect on the trajectory of the conversation.

To conclude, change is the evolution of new meaning through dialogue. Ultimately, stakeholder focus groups are not only a place to discover the pertinent issues of stakeholders, they also provide a chance to see individuals change in response to interacting with others. This individual change, in turn, creates a change in the stakeholder dialogue that provides insight toward innovation and policy development. Ultimately, focus group facilitation should create a safe environment that allows the beliefs of stakeholders to be gently challenged.

SYSTEMS THERAPY: A METHOD FOR MANAGING HUMAN SYSTEMS

Focus groups are intended to bring people together for discussion. Once the group is formed, a human system is created. This human system can function but frequently gets bogged down by hidden agendas and political interests. As previously discussed, paying attention to cognitive and social factors such as learning and change will significantly heighten the success of a focus group. This special attention requires a certain kind of management of the group relationships, interaction and information acquisition. The management encourages individual creativity and expression while maintaining focus on the topic of interest. Successful facilitators tend to have a rule-of-thumb understanding of human behaviour, are familiar with a set of group problem-solving methods and have a natural charisma. These skills are important and generally adequate for many problem situations, but as problem complexity and interpersonal strife increase, basic facilitation skills are insufficient. An issue such as pest management requires facilitation by people who can find their way through individual, social and cultural complexity.

Theoretically driven facilitation provides a balance to the overemphasis typically placed on focus-group process and methodological 'tricks'. A good sociologist or anthropologist can sufficiently analyse acquired data from a focus group. A developmental psychologist can identify the cognitive changes made within each participant of the focus group. But it is optimal if all of this can be combined in one person who can also comfortably manoeuvre within human systems and has therapeutic skills when the human system becomes paralyzed. In response, we introduce a technique called systems therapy.

Systems therapy is a skilled craft useful for discovering meaning and providing interventive and management strategies that improve the health of human systems which have organized themselves around a problem. Systems therapy is a unique hybrid of four distinct disciplines: cultural anthropology, developmental psychology, family therapy and systems design. First, cultural anthropology (Veatch, 1969; Husserl, 1970; Radin, 1966; Lofland, 1971; Miles and Huberman, 1994) enables the elicitation, analysis and organization of the socio-cultural information belonging to stakeholders. Second, developmental psychology (Piaget, 1952, 1955; Erikson, 1963, 1964; Kohlberg, 1969, 1973; Baltes et al, 1980) provides theoretical insight into the dynamic processes of how human beings develop over time. Third, family therapy (Watzlawick, Beavin and Jackson, 1967; Lederer and Jackson, 1968; Satir 1964; Haley, 1963; Becvar and Becvar, 1988) provides a means of eliciting information amidst a conflictual environment, fresh descriptions that energize dialogue into new meaning, and interventive strategies that increase healthy functioning of human systems. Fourth, systems design

(Woods, 1993; Wymore, 1993; SEI, 1993; Shenhar, 1991; Flood and Jackson, 1991) contributes methods of managing technical information of vast complexity into precise solutions with a high standard of accountability. Systems design, the fourth element of systems therapy, will not be discussed in this chapter.

Overall, systems therapy uses strategies that assist human systems to get *unstuck* from their present dilemmas while simultaneously documenting and managing relevant information. People tend to become stuck in a pattern of dialogue simply by attempting the same solution to their problem over and over again. Solutions may have worked once before and continue to be applied; however, although the attempted solutions are repackaged or redesigned and look different, they are inherently the same. Here, an adherence to these attempted solutions becomes the problem. An approach such as systems therapy is invaluable for socially complex problems since stakeholders are generally unaware of the degree to which their dialogue with others hinders or enhances progress. In addition to gaining a richer understanding of a system that improves product and service delivery, the process of systems therapy is useful for most situations where humans interact (focus groups, interdisciplinary teams, management struggles).

To enable a system to become unstuck, systems therapy first defines the problem and the problem space. Second, it gains an understanding of the present system that includes living and nonliving objects and the complex interaction occurring between and within the objects (Mayhew and Alessi, 1995). Third, systems therapy develops interventive strategies by refashioning the stakeholder information. The intervention is built on identified knowledge that already exists in the human system, but this knowledge is refashioned in a way that perhaps has not yet been considered. This introduction of new information to the group challenges the present paradigms, introduces new ways of thinking and jars the system into new realities. In other words, the intervention creates change within the human system.

Although a detailed description of systems therapy falls outside the scope of this chapter, how systems therapy differs from typical focus-group methods is demonstrated in the following example. Prior to discussing this study, we present a result obtained from using systems therapy. A dualistic debate pattern was identified which tended to confound and stifle the stakeholder dialogue. The systems therapist uses this knowledge to 'refashion' the dialogue in an attempt to intervene and help improve the stakeholder debate.

Dualism: Maintaining the 'stuck' system

Debate is defined as discussion, dispute, consideration or deliberation where *both* sides are presented (Random House College Dictionary, 1984; italics not included). The word 'both' conjures images of two opposing sides. The debate within the issue of pest management also appears to have a dualistic appearance: those who support the use of chemicals and those who are against it. Obviously, the discussion is not that simple. Stakeholders line up all along the spectrum. One state legislator said: 'I think there's a continuum. There's a place for the prevention and there's a place for, at some point, the inputs.' When the discussion heats up, people are either driven or are perceived to uphold extreme positions. When asked about this polarism, an entomologist said: 'I think people get pigeon-holed. If I'm talking, if I even mention chemicals, maybe I'm over here [gestures to the right], you're putting me clear over here on this end down here [gestures to the left].' A farmer resented being placed in an extreme pro-chemical light when he said: 'I don't think any farmer wants to use anything that he thinks is going to hurt the environment whether it's chemicals, fertilizer, too much manure, or any product.' When top executives of a chemical manufacturing company were asked if they thought the pesticide debate had become dualized, one reluctantly replied: 'That's probably what it comes down to, you know.'

These statements document how the dualistic nature of the debate has maintained an *us versus them* type of interaction that prevents further progress. The dualistic view is not without its function to individuals since it enables a person to compartmentalize a complex issue into two sides that guide his or her thoughts and subsequent speech. Although an individual's cognitive tension is relieved, the overall discussion is constricted and reduced to a dualized debate. The systems therapist searches for new descriptions that can be used to intervene and move the focus-group process forward.

Triangles: Potential source for interventive strategies

Although people tend to compartmentalize problems into two sides of their thinking and speech, Bowen (1976) argues that the smallest and most stable social system is the triangle. The triangle has definite relational patterns. A triangle consists of a comfortable and close relationship between two people with a less comfortable outsider. This two-person system, or dyad, forms a coalition at the expense of the third person (Minuchin, 1974). The dyad seeks to preserve the close relationship while the third person seeks to form a togetherness with either person of the dyad. The most common strategy utilized by the third person to get into the relationship is to attack either member of the dyad. This is attempted by many means such as speaking pejoratively of an other's competence. These attempts by the third person threaten the closeness of the people in the dyad; this, in turn, only strengthens their closeness and further distances them from the third person. An illustration of this phenomenon within the stakeholder discussion on pest management will be discussed later in this chapter.

Social systems contain innumerable interlocking triangles typically invisible to their members. The previous example demonstrates how triangles can be observed in the pesticide debate and how the systems therapist can reintroduce, or *reframe*, the dynamic of a system. This action opens space for people to see the situation differently. In turn, this lessens the stranglehold on the system and allows change to occur. Further application of this phenomenon will be discussed in the results section.

Stakeholder Analysis of Pest Management

This section describes the specific methods used in a study of stakeholder discussions on pest management. The analysis contains stakeholder viewpoints and human interaction patterns among stakeholders in Iowa as well as corporate executives of a major chemical manufacturing company. The purpose of the analysis was driven by two potentially rich and highly complex interests:

- to gain an understanding of the human phenomenon that occurs when stakeholders of pest management interact;
- to discover new business opportunities for pest management.

Stakeholder groups

The first step in this study was to select representatives who have a stake in pest management. Table 6.1 displays the kinds of stakeholders selected for each group. This selection was based on a pilot study of two focus groups conducted by Mick Mayhew and Bill Vorley in Iowa. Each group consisted of farmers and crop consultants. In this pilot study, the groups that had a stake in pest management were evident. The participants of the pilot focus groups expressed feelings of being misunderstood and projected blame for the misunderstandings onto other people groups. For example, they identified media portrayals of frightening scenarios caused by using pesticides (for instance, Alar). These scenarios were allegedly started by environmentalists who also received the attention of politicians. In this example, four stakeholder groups readily become apparent: media, consumers, environmentalists and policy-makers.

Since the pilot study identified that most consumers have little to no knowledge about farming, we selected urban consumers that were at least two generations removed from the farm to participate in the stakeholder groups. We began to see that some of the farmers' concerns were contingent on more information – from economists, biologists and social scientists – so these were included as stakeholders. A discussion of potential markets emerged in the pilot study, so grocers were also identified as a stakeholder. Farmers are influenced by the education and information they receive. Consequently, crop consultants and extension agents were identified.

Chemical manufacturers are an obvious stakeholder in the pesticide discussion. We were also interested in gaining the perspective of corporate executives from the chemical industry. Interestingly, a few manufacturing companies expressed interest in the study. We selected one company and obtained the participation of four of their executives (see Table 6.2).

Participation was elicited of individuals who represented the interests of each stakeholder group. Obviously, one or two people cannot represent the broad spectrum of ideologies inherent in each stakeholder group, so some stakeholder groups were subdivided. Farming was divided by those that farm conventionally by using chemicals and those that farm organically. Grocers were divided into large chain grocers and organic grocers. The majority of focus-group participants were male with minimal female representation. This certainly was not intended, but the names of potential stakeholders that we were able to acquire were predominantly male. Most of the women that we sought from the pool of names were unavailable to participate for legitimate reasons. In fact, all of the people originally contacted were eager to participate, but final selection was based on schedule availability. Time constraints caused us to press on and, consequently, stakeholder representation became the primary focus of participant elicitation.

We noted during the study that two stakeholder groups had not been considered – bankers and corporate stockholders – but it was too late for their inclusion. Bankers were considered stingy keepers of the purse strings by farmers wanting to try unconventional practices. Executives of a chemical manufacturing company identified the desire for profit by corporate stockholders as a limiting factor. It would have been extremely valid to add the thoughts and perspectives of bankers and stockholders to the stakeholder dialogue.

Stakeholder session format

Figure 6.1 depicts the overall process of meetings with stakeholder groups. Each group met for three 90-minute sessions (approximately 3.5 weeks apart) over an eight-week

Table 6.1 *Stakeholder participants in each Iowa focus group*

Group A	*Group B*
Agricultural economist	Grocery chain retailer
Conservationist	Conventional farmer
Conventional farmer	Entomologist
County legislator	Media reporter
Crop and soil consultant	Conservationist
Educator in extension	Educator in extension
Geological scientist	State legislator
Media reporter	Chemical manufacturer
Organic farmer	Sociologist
Organic grocery retailer	Organic farmer
Urban-dwelling consumer	Urban-dwelling consumer

period. Group A consisted of 11 people who lived in or around a city in eastern Iowa. Group B included 11 people from a city in central Iowa and the surrounding area. Group A met in an Extension office and Group B met in the Marriage and Family Therapy (MFT) Clinic on the campus of Iowa State University.

The MFT clinic was selected as a site for Group B to enable a third group, the reflecting team (Table 6.3), to observe the interaction of the group. The reflecting team observed the sessions from behind a one-way mirror of an adjacent room. This team of six people, all holding doctoral degrees, were selected for their knowledge of pest management or their knowledge of human interaction and process. They were asked to observe the session of Group B and subsequently comment on the process for 30 minutes while Group B observed the discussion of the reflecting team. The purpose of the reflecting team was to provide an additional vantage point to observe the pesticide dialogue and also to introduce new information which might influence Group B. Group A did not have a Reflecting Team but were exposed to comments by the Reflecting Team about Group B.

A two-hour focus group with the corporate executives occurred in week seven after the second and before the third stakeholder focus group sessions. The four executives (see Table 6.2) were first asked to react to the preliminary results from the first and second stakeholder sessions. Some preprepared questions were also introduced for their reactions. The second and third sessions began with a brief summary of the previous session. The group members were asked to respond to the accuracy of the summary and corrections were elicited. This *member check* maintains the integrity of the stakeholders' views. This corrects the trajectory of the ongoing analysis and heightens the precision for a final document.

Table 6.2 *Focus group participants of top-ranking executives of a major chemical manufacturing company*

Vice-president of disease control business unit
Vice-president of planning
Vice-president of environmental public affairs
Vice-president of research and development, crop protection

Documentation

Documenting events that occur in a stakeholder analysis is essential (Morgan, 1988, 1993). The analyst attempts to develop ideas that are theoretically grounded in the data (Broadhead, 1983; Glaser and Strauss, 1967, 1970; Strauss, 1987, 1991; Strauss and Corbin, 1990; Weiner, 1981) that can ultimately be traced back to the data (Mayhew and Alessi, 1994). Documentation can take many forms and levels of rigour. Flip charts and note-taking are typically the means used to document the discussions. The problem with these methods is that most of the dialogue, language and terms become lost, leaving a simple distillation of thoughts. These methods obviously serve their purpose when only a simple analysis is required.

The method of our choosing seeks to capture *everything* that occurs. The stakeholder focus-group sessions were audio tape-recorded, transcribed and all identifiable information (for example, names, employers, city of residence) pertaining to the stakeholders was removed. Additionally, the analyst maintains notes taken during and immediately after each session and throughout the analysis process. This critical and methodical note-taking process is called the *audit trail.*

Analysis

The transcripts are analysed for narrative (Bruner, 1990; Cronon, 1992; Riessman, 1993; Schafer, 1992; see also Bakhtin, 1981, 1986) and content (Krippendorff, 1980; Namenwirth and Weber, 1987; Stone et al, 1966; Weber, 1990). An ethnographic approach is used to discover, identify and coordinate the units of socio-cultural knowledge held within and between stakeholders (Bernard, 1994; Denzin and Lincoln, 1994; Guba and Lincoln, 1989; Lincoln and Guba, 1985; Spradley, 1979). A thick description emerges (Geertz, 1973) about the many layers of social and linguistic phenomena that simultaneously occur during the session. For instance, there are the actual words that are being communicated on one level, the personal posturing that occurs on another level, and the alliances and coalitions which are being established at still another level. A taxonomy of themes emerges from the final analysis of transcripts and audit trail. This is as far as we took the present analysis and a sample of these themes will be presented in the next section.

If this study were to be brought to the next phase of systems therapy, precise requirements for future designs of product or service delivery would be derived from the themes. After the themes are identified from the transcripts, the transcripts would further be subjected to content analysis. This would provide a means to apply empirical and statistical methods to the textual material (for instance, transcripts, notes). Once the innovations are designed, reactions from the stakeholders would be elicited to test for accuracy. The content analysis allows for testing the new design against the requirements.

The Results: The Dialogue Unravelled

By now the reader has been acquainted with how human systems get stuck around difficult topics such as pest management. This is typically the result of selective stakeholder involvement early in the design phase of product or service development. At the very least, the system is maintained by individual and social reluctance to change.

At this point, three resultant themes from the study will be discussed:

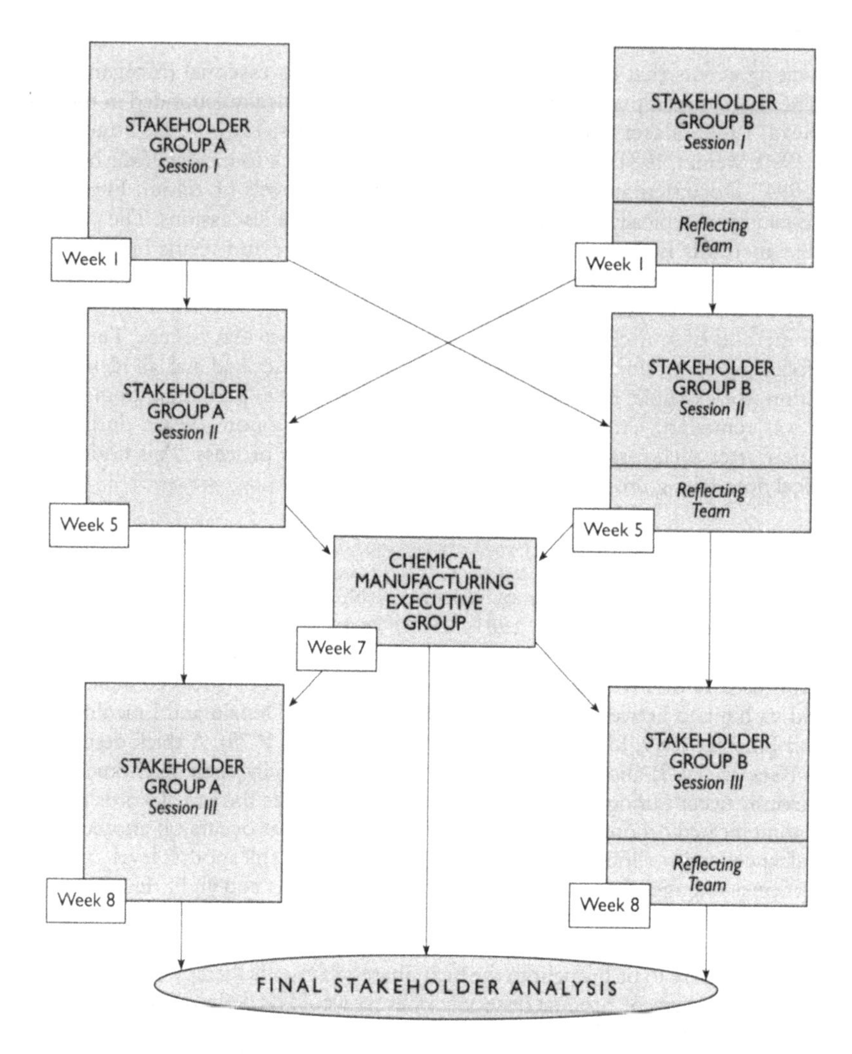

The boxes depict the consecutive order (from top to bottom) of the stakeholder focus group sessions over an eight-week period. The vertical arrows depict the feedback summary of the previous session to each group. The diagonal arrows depict the cross-fertilization of ideas between groups.

Figure 6.1 *Display of the process used in the stakeholder analysis of pest management*

- the 'triple–S' debate: system, science, sustenance;
- a systemic view; and
- a question of the head or the heart: education as solution?

Each of the three themes will be described by remarks from the stakeholders. These remarks are direct quotes from the participants and are identified by the respective stakeholder group that they represent. The only edits made to the remarks are illustrated in brackets. These edits were made to protect the confidentiality of the participants.

Table 6.3 *Reflecting team that observed and commented on stakeholder group B*

Agricultural technician
Agronomist
Futurist
Marriage and family therapists
Sociologist

The 'Triple-S' debate: System, science, sustenance

Inconsistencies of opinion, arising from changes of circumstances, are often justifiable.

Daniel Webster, *Speech*, vol v, p 187

As previously discussed, the toxic debate of pest management is centred on the use of chemicals. Upon careful scrutiny of the stakeholder dialogue, there are not merely two sides to the dialogue, but three. There exists a triangle of stakeholder groups that have gathered around the pest management issue: system, science and sustenance. A discussion of the three groups follows, and highlights how they are uniquely different based on the stakeholder participation, merit system and focal point (see Table 6.4).

The system

We have all experienced the feeling of powerlessness at the hands of some ominous, faceless entity. The culprit of disempowerment is often ascribed a one-size-fits-all alias: the system. The system is a jargon word inherent within an environment that believes that there are the have's and the have-not's. The regime has the money and makes the decisions; it is the seedbed of power and control. In moments of sheepish glory, an exacted revenge is to hurl derogatory statements about the system but to hold no illusion of actually hitting the target.

The system made more than a cameo appearance within the stakeholder discussion of pest management. Representing the system were large corporations (manufacturers and farms), banks and government. An organic grocery retailer said:

But it just seems to me that around the corner may be the opportunity for every farmer who's now working the land to work as just a manager. His parents might have owned the land but now it's owned by some giant international corporation and believe me, they are not worried about soul, spirit or anything else and they'll grow organic or commercial, depending on whatever brings in the very best buck.

Table 6.4 *The Triple-S debate*

	System	*Science*	*Sustenance*
People group	Banks, corporations, corporate farmers, government	Scientists	Conservationists, environmentalists, organic farmers
Merit system	Money	Discovery	Sustenance of human and natural resources
Focal point	Short-sighted	Hind-sighted	Far-sighted

The following exchange between a policy-maker and an organic farmer occurred:

Policy	*There is no such thing as a family farm anymore. They are corporations, they're businesses...*
Organic farmer	*But the system you are talking about is supported by the legislature and Iowa State University research...*
Policy	*...It's a reaction to the forces that are going on, and it's concentration, it's economics ...*
Organic farmer	*I have a good friend who...is having a devil of a time getting a farm loan to get back on his feet. He pasture/farrows hogs. Bankers do not believe in pasture/farrowing hogs.*

An interesting phenomenon regarding the system occurred. When a stakeholder who was identified as a member of the system displayed honest vulnerability, he no longer was treated as an enemy. For instance, the above discussion between the policy-maker and the organic farmer occurred after the conventional farmer's statement that, on average, he needs to earn $40,000 annually for his family to live. Prior to that, the conventional farming practices of pest management were defined as the problem, due to their use of chemicals. Once he stated this, it sent some stakeholders, and ultimately the dialogue, into disequilibrium until the stakeholders could establish a new scapegoat. It was at this point that the system emerged as the entity maintaining the dysfunctional practices of pest management. The farmer was seen as a hostage to the present practices. This phenomenon exemplifies how human systems have linguistic mechanisms to equilibrate and self-correct.

The vantage point of the system tends to be short-sighted – approximately five years. A conventional farmer said:

> *But personally I don't go much more than a five-year plan. You have a general idea, you know, I might want to have more acres ten years from now or you have a son coming up or something like that. To me it's changing so fast that I don't try to lock anything...*

A policy-maker in the state legislature expressed frustration about the short-sightedness:

> *Speaking from the political process, you know we're talking in most cases in two-year terms. So most legislators think in two years... We don't think, I mean we as a society, especially in the United States, we don't think long term. We think in just short little blocks of time and you can't think and anticipate what these issues are gonna be and how to solve them...And we just don't have time, we don't take the time, maybe that's a better way, in the legislature, in the political arena, to look at the big picture and decide what the big picture is going to be and how to deal with it...But it's not going to happen and I'm feeling a little pessimistic about it.*

The measurement of merit used within the system is money – money is the motivating force. An organic grocer reported that the system maintains current practices with money:

> *And if you go to the bank and you say, I've got this great idea. I've got a vision. I'm hard working. I have a plan, a study. I even have people I can consult with...and the bankers just send them outta here. You're a weirdo, we have no interest in loaning you any money. Without credit you can't produce.*

A planning executive within a chemical manufacturing corporation stated that, although scientific guards are used to ensure safety, in the end, money is the key player:

> *An [executive of environmental public affairs] said, we do the science and we make sure what we put in the marketplace is not going to result in harm. But that's not our driving force. Our driving force is return to our stockholders in the end.*

Ultimately, the general public is left without an assurance that their best interests are protected in the system's drive to make money. An urban-dwelling consumer was probed by the systems therapist on this issue:

Consumer	*As a consumer, I think I must watch too many Twenty-twenty's because nothing surprises me and I almost expect that even though there is all this control and regulation that if the money's big enough or the pressure is great enough or whoever might have the money, that it doesn't matter how many controls or regulations we have. So my point is my trust is very minimal as a consumer.*
Therapist	*Because...?*
Consumer	*Because of money again and that there's probably someone, and I don't know who that would be, within the system, if it's the corporate giant that has the product they have to get through or get back on the market, or change the label, whatever it might be, to get it to happen for the wrong reasons, then it will happen. It just probably hasn't become public yet. So I think trust is very minimal. I think probably control and regulation from my perception as a consumer is probably overdone but I don't know how those two come together because I think greed and money is still the driving force. So I don't have a lot of trust or confidence in the system.*

Summarily, the voice of the system is comprised of corporations, government and banks. The system maintains a short-term view with an eye on profitability. There is a wide margin of distrust of the system but people appear fatalistic that nothing can be done about it.

Science

The role of science emerged as another significant voice within the stakeholder dialogue. Science serves as a source of information, as one crop consultant stated: 'We work closely with Iowa State Extension. They give us a lot of information as far as research on current issues.' Science was quickly called into question as being too focused by a conservationist:

> *I would like to relate to what [the organic farmer] has said there because I think when we have a problem, we have a tendency to move in on that particular problem and science focuses on it and tries to cure it. I think, and in doing so, we have become so specialized in hogs, or in soybeans, and we have developed many monocultures out there based on agricultural practices, the science and economics behind it, for the sake of efficiency.*

At a later point, a scientist who works for a large chemical manufacturer warned that decisions need to be based on nothing less than science:

> *And to start banning pesticides or to take some of these tools out of the box for political or emotional issues, and not based on science, down the road, I feel, there is going to be a very limiting factor to maintaining the way of life you've got.*

In fact, he confidently went on to say: 'We got the science, we got the science down to fact. What we need is better communication.' On the other hand, he later reported that it is difficult for science to give finite answers because of the complexity:

> *I think in the context that [the legislator] brings up, in the fact that she wants a bottom line answer and she wants it in black and white. Is it a yes or no? Can I vote yea or nay? And I think it's with everything in science, there's no black and white...And it's tough for [the biological scientist] to go ahead and make a recommendation and write it down in black and white on an eight-and-a-half by eleven page and say okay, here's what you ought to do...You know, there is no black and white. It's a biological moving system.*

An agricultural technologist of the reflecting team observed:

> *There was a lot of certainty about the safety of chemicals. So, if it is true that the science that produced that result is the same science that can't grapple with systemic problems, what is the answer? It leads to skepticism. In one case you're very certain and in another case you're very uncertain when science is being used in both cases.*

The credibility of science was called into question by a conservationist because, as he suggested, science is being held hostage by the system:

> *I think that you have to look at the scientist that comes out of a corporate structure or you have to look at the scientists we have doing the research here in a public institution. I think that flow of information should be free, but the time doesn't exist, the money isn't allocated that way...I feel our scientists have been sort of hamstringed. In other words, you go out on a project and this is your baby and you stick on that. This comes from the corporate side.*

An entomologist responded:

> *There is tremendous pressure for us to get outside funding for one thing...the proposals to get outside funding have to be very focused...so you end up with a lot of small goals. So that is one thing, and then...the drive for refereed publications in scientific journals. Some very practical information is probably not publishable in scientific journals. So there are two things that are driving the isolation and I agree with you, it's here.*

Another attempt to bring science into perspective was made by a rural sociologist who questioned if society's view of science has become part of the problem rather than just part of the solution:

> *I wonder, and I'm not an expert in this area, but I wonder if a lot of the pest problems are not really human-created problems. I think I would like to explore the possibility that with our fascination with technology and big science, and in fact, if we have not created about as many problems as we have solved.*

In sum, the role of science is to discover new truth. This is done by designing an experiment, executing it and publishing the results in a journal. The merit system of science is discovery and it uses the amount and weightiness of journal publications as the standard of measurement. Although science studies phenomena that can provide implications for the future, its standard of merit is based on historic accounts. Science was observed as hindsighted and specific and, therefore, unable to answer many questions that are useful to the degree that policy-makers, farmers and the general public would like them to be.

Sustenance

The critical factor for this group of stakeholders is simply to sustain life. Stakeholders desire a means of agriculture that will ensure the health and safety of natural resources such as soil and water. They also express a desire to secure the health of people – consumers and workers – now and for future generations. This is illustrated in the words of a conservationist, urban-dweller, grocery retailer and an environmentalist:

Conservationist — *So I don't think our land is being used in the country in a manner in which it will sustain itself or sustain the food for ourselves.*

Consumer — *From the consumer side the big thing would be safety, as far as environmental hazards, and also the physical effects that it would have on consumers, which opens up a whole can of worms, I think, as far as long-term effects and short-term effects.*

Retailer — *We need to look for a system which protected the health of the consumers as well as one that did not pose a hazard to the workers in the field, applicators, or the actual farmworker or person who is actually harvesting the crops. It would also protect the groundwater and help the soil condition and so forth. So those would all be the factors that would be most important to me.*

Environmentalist — *You have to think in terms of the soil. A hundred and fifty years of farming and two-thirds of the black soil is gone. What happens in another 50 years, which may be some of your lifetimes yet, and it could be the rest of it could be gone...I don't have a snappy answer but I do have a, I have a stake in what farmers do. I have a very definite survival stake in their survival. If they don't grow the food, I die. If they poison the water, I die...So I have a stake in them growing food without poisoning the water. Now, as a taxpayer, let's say I'm willing to pay for that. Grow the food without poisoning the water. I'm willing to pay you for it...Because what it would mean is, if he could earn a living, if he works to save the water and the soil, which is in my best interest and I'm willing to pay him for it, collectively we should maybe all pay for it, and he's working at that and he grows the food and he saves it and he gets $100,000 cheque from our taxes because he did a good job with the soil and water, I think that's great! Because 150 years down the road, my grandchildren are still going to be able to survive.*

The view of sustenance tends to be far-sighted. Although sustenance has a concern for the present safety of human and natural resources, it sees the gradual deterioration of resources as threatening the future of life as we know it. Listen in to the comments made by an environmentalist who offers a perspective of more than five years. Also listen to the cold splash from the geologist; while merely intending to make a humorous quip, he displays the profundity of short-sightedness:

Environmentalist — *We've only been farming as a species for about 10,000 years. Maybe there hasn't been time to evolve. You know, the hunter/gatherers had several*

	million years to develop the bow and arrow and finally the gun. You know they were able to do it much quicker, more efficiently. But maybe we haven't evolved to the point that we can look ahead. That we can actually make plans for more than five years. If our species isn't capable of doing that. If we had a leader rise up and say, folks, we have to think about the oil depletion. You know, who would listen to him?
Geologist	*That's what you call one-term presidents! (laughter)*

Although our society seems readily able to dismiss the far-sighted amongst us, the effects of short-sighted practices are not quickly remedied. An interaction between the geologist and a crop consultant displays the poignancy:

Geologist	*Wherever you have water that can be affected. If it has precipitation in it. If it has water that fell as precipitation in the last five decades, it has nitrate in it. And before, the only time it did was because the wells were lined with brick and there was a hog lot or a cattle lot right there. Now it is because of application of fertilizers over the last 40 years. And where you have the data, the two just climb together. (long pause)*
Consultant	*Is it still rising or has it tapered off or is it going down?*
Geologist	*I would say the best information that I have seen suggests that nitrogen fertilizer applications have declined about 15 per cent on a per acre basis in Iowa in the last six years. But with the kind of climatic variability we see, we're not going to see a change. What we have probably seen there is that conditions will not get any worse. It will slowly start to improve. Which is all anyone ever expected. Cures don't happen overnight.*

In the final analysis, the focal point of sustenance is far-sightedness and its merit system simply is *life*. Although individuals within system and science argued that life is also a concern of theirs, it was noted that the primary measure of merit within system and science is money and discovery, respectively.

Interaction effects

How do system, science and sustenance interact within the stakeholder dialogue? There is an apparent coalition between system and science with sustenance as the excluded third party. In fact, this excerpt by a grocer depicts the suspicion of system and science as being in cahoots with each other:

> *...because obviously you can see right off the bat that the system, the banks, the large, powerful, well-financed chemical companies are funding the science and you can darn well bet they're going to give them what they paid for in many cases. Now I know that's going to offend a bunch of scientists. Many feel that they are genuinely researching, you know, scientific issues of merit in themselves, but I think that in many cases science is right there standing right behind, doing what the chemical company or the bank or whoever was going to be the bad guy, doing whatever they tell them to do and they're going to make it look good and put the right spin on it and stand up with the letters after their name. So, I'm a little, I mean I know there are scientists who are genuinely researching issues, but I'm a little suspicious of science being a handmaiden to the money.*

Although this sceptical belief was introduced and argued in both groups, it represents the kind of beliefs – true or not – that the systems therapist encounters and must treat as 'reality', because it is reality to some.

Science took a lot of hits in this study. It holds a great deal of prestige in our society and typically provides the 'official language' for settling disputes. Science is used to ensure safety of agricultural practices before corporations can put an innovation into production. When future projections of the innovation's safety are called into question, it is incumbent on the accuser to 'prove it' by using scientific means. Obviously, the accuser is unable to provide evidence for the argument because there exists no data from the future. As science is called to settle the dispute, quality of life issues espoused by sustenance are acknowledged but usually dismissed.

The criticisms that sustenance makes of science beg the question that perhaps science operates within different levels of analysis. In other words, is it befitting for science to serve as primary consultant to those making comprehensive decisions about the future – amidst all the complexity – based on extremely specific empirical past data? Sustenance is not willing to give this high-level trust to science as noted in this interaction between an entomologist and an environment-friendly legislator:

Legislator	*I think what we're doing, it's a fundamental difference of philosophy. And [the conservationist] and I and [the organic farmer] are sitting on the bottom line. It's a fundamental difference of philosophy in terms of do you trust science and technology and chemicals or do you not. And that's a bottom-line difference in philosophy...*
Entomologist	*Yeah, that's probably about right.*
Legislator	*...and I'm real sceptical.*

Systems therapy intervention: a reassembled Triple–S

So what role should science play in the pest management dialogue? Should system play a different role? Is there another role that sustenance can play? To answer these questions, one can begin by investigating the results of the various alliances. Figure 6.2 suggests that the main result of the system–science relationship is technology. An entomologist clarified the distinction between science as discovery and technology, and as the application of knowledge:

> *Science and technology are really two different things...The goal of technology is application so, you know, there's more there than just discovering some facts. Scientists are also looking to why those things are and saying how can I use this knowledge, this basic knowledge, and learn to apply it...Science is thought, not action, and technology is action. Technology is taking what we know about the natural world and applying that.*

A member of the media stated that technology is closely tied to system: 'I think, in a way, technology is one of the forces that makes that system of money work.' A biological scientist reminded us that technology has made some very remarkable contributions to humanity:

> *We are modifying a natural system in order to produce something. And we're never going to have to not continually have some kind of inputs into that system in order to make it work because we're modifying nature. That's what technology is.*

But others are not as impressed. An organic farmer quipped:

> *You said something before, you said science is thought and not action and that technology was action. And, I'll be honest, my natural finish of that statement is, 'and not thought' (laughter). I'm serious, that's what I'm getting at here, which is, that we tend to use technology without looking at the broader implications, which is the problem I have with technology. Not that it's bad. It's that we don't look at it within the broader system, which is what we have to do to make it make sense. So that's the problem I have with it.*

What would happen if an alliance was struck between system and sustenance? Could system reap some of the same profitable results as it does from their union with science? How would this affect science? These questions were presented to the focus groups and were received favourably, as indicated by one of the urban-dwellers who, incidentally, is an engineer with an international corporation:

> *I think it's a good idea, but you have to come on some kind of common ground. I believe at this point, the common ground may be science. Because I heard a lot of good things from the environmental standpoint that I thought was very scientific. It tends to be a lot of where I would lean toward as far as issues of studying the amount of erosion or the depletion of the soil, the topsoil, the vitamin issue that [the grocery retailer] brought up. A lot of these other things, to me, are very scientific. They carry a lot of weight. It just depends on what you want to look at. Both of those have to let go of their ideologies. You know, the system has to be willing to accept the other side and both of them have to accept one another.*

Notice how system and sustenance are personified into entities that have to find ways to get along. Meanwhile, science is positioned as an excellent *source* of information, though removed from the primary relational interaction. Figure 6.3 depicts this new arrangement with the solid line between system and sustenance representing a primary relationship and the dashed line representing a secondary relationship: that of a consultant. This new arrangement opens space for information flows to emerge that can serve to stimulate innovative thinking.

Members of the chemical-manufacturing company focus group positively entertained the idea of a system–sustenance partnership to generate new pest management options. These executives included the vice presidents of the disease-control business unit, environmental public affairs, research and development of crop protection, and planning:

Disease	*But this group would be interesting just to learn from. Like, if you were developing a product...*
Planning	*What would you develop?*
R&D	*Well that's basically the opening question.*
Disease	*And it would be interesting to see what they would design.*
Planning	*Well, the interesting thing about what you're doing is they learn more about each other's needs and understand each other, they start changing the way they view things.*
Disease	*Yeah, that would be interesting.*
Planning	*It's gotta be.*
Public	*But we do a lot of work with growers.*
Planning	*We don't do it formally. We don't go out with the notion of a product concept, of using a group to develop a product concept. We have our product concept...*

Public — *We do with growers.*

Disease — *Oh no, with growers? We do with growers.*

Planning — *What do you do?*

Public — *We go out with three or four different options. You can't go out and say, if you were to do this, but we go out with three or four different options.*

R&D — *Would this be attractive to you?*

Disease — *Yeah.*

Planning — *But see, that's exactly the opposite of what I'm saying. You go out with the idea. What I'm saying is let the idea come from ...*

Disease — *Well, we get that from 'em, but if you do that, you're there for usually a week. So we don't do that.*

Planning — *Yeah, you're right, we don't.*

What prevents these executives from actually talking to the growers, learning from them, allowing new ideas to emerge without first priming the discussion with a few options that probably inhibits creativity? A week's worth of time? What might be the benefits? One of the policy-makers was also intrigued by the idea:

> *There is a potential to find some common ground too, which is important, because, you know, we're talking about the polarization on one side here and the other side, but there is common ground that we can agree on and start from there. And I'll tell you, there is, like, this new revolution that is happening because I'm involved in three projects where this is happening. And it's pretty phenomenal.*

One of the greatest constraints to this process is jarring people loose from their way of thinking. While some can grasp new ideas by assuming a curious stance, others have difficulty seeing beyond their own experience, in spite of the questions by the systems therapist:

R&D — *But you come back to the environmental groups which we've had difficulty in talking with...*

Public — *That particular group won't even return our phone calls. We've tried. They don't. I don't want to spend a lot of my time trying to convince some of those people that they're wrong or that they don't understand the nature of the business. When their livelihood depends on them bashing our business, regardless of how much information they do or do not have.*

Therapist — *But the way you just worded that...*

Public — *It's a waste of time.*

Therapist — *But, what you said was, I don't want to waste my time educating them on what's really right.*

Public — *Or trying to. We're trying to.*

Therapist — *Do you think that's maybe what keeps them from wanting to get together because that's what they expect?*

R&D — *In other words, it's not a give and take.*

Therapist — *You're calling to talk to me, you're calling to convince me. So do you reckon that...?*

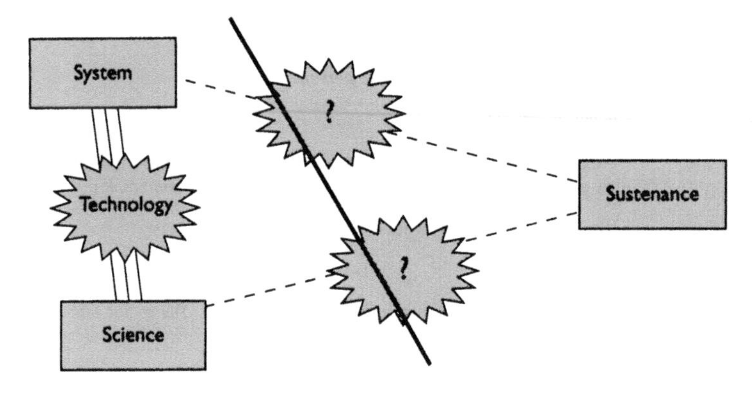

Figure 6.2 *Offspring of various unions within the pest management dialogue*

Public	*Is that bad?*
Therapist	*Well, I don't know. They don't call you back.*

The reactions to the idea of a system sustenance alliance by the focus-group members ranged from enthusiasm to guarded interest. The only exception was a rural sociologist who stated:

> *Well, it's nice to say look for win–win solutions or look for the importance of coalitions and many people argue that coalitions will become more important than special interest. You know, that's a 20-year-old sort of literature in there and, quite honestly, it's pretty hard to find very many win–wins, because win–win solutions require compromise and many of the special interest groups, advocacy groups, have really become social movements.*

Systems therapy does not seek to establish win–wins, compromise or even solutions to the interactive problems. It merely seeks to understand the stuck system, introduce new information through questions, comments or strategic interventions that will elicit new knowledge from the stakeholders and the inherent dynamic processes. Solutions are derived by an innovator's creativity that is greatly enhanced by this new knowledge.

The systemic view

> *So naturalists observe, a flea has smaller fleas that on him prey;*
> *And these have smaller still to bite 'em, and so proceed ad infinitum.*
> *And the great fleas themselves, in turn, have greater fleas to go on;*
> *While these again have greater still, and greater still, and so on.*
>
> Jonathan Swift, *Poetry, A Rhapsody*

The word 'system' first appeared in 1603, but the currently celebrated meaning was developed by Norbert Weiner (1948) with people of many disciplines (mathematics, medicine, physiology, psychology, anthropology, economics, anatomy, engineering, neurophysiology, sociology). Inspired by his work with anti-aircraft guns during the Second World War, Weiner developed a theory that compared machines with living organisms. This was investigated to provide an understanding and control of complex

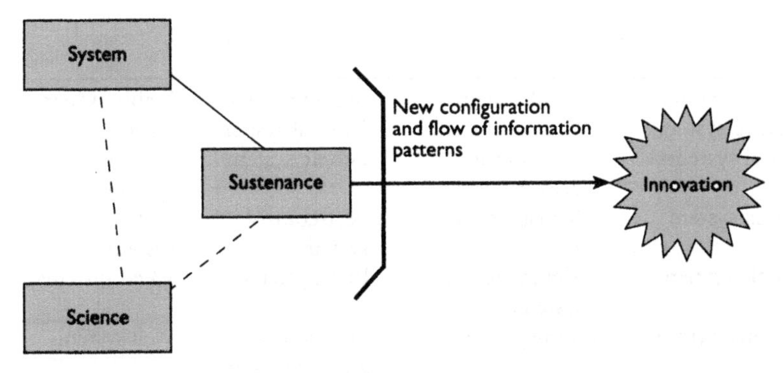

The solid line between system and sustenance represents the primary relationship and the dashed line represents a consulting relationship. This new arrangement creates new information flows and creates a potential for innovations to emerge.

Figure 6.3 *Possible restructure of Triple–S*

systems. Ludwig von Bertalanffy (1968) published similar work from Vienna, based on general systems theory in the 1950s.

The concept of systems fashionably permeates much of contemporary dialogue in academic and professional circles. The stakeholder focus groups were no exception. The word 'system' or some derivation of it was used 335 times throughout the entire study. Interestingly, the word was used in numerous situations that were grouped into seven categories depicted in Tables 6.5 to 6.11. Table 6.5 displays the different applications of the word as mentioned by the stakeholders throughout the focus-group discussions. According to the stakeholder analysis, systems come in all shapes, sizes and scales (see Table 6.6). Table 6.7 shows that systems are alive, adaptive and can reach outside of themselves. A heavy emphasis was placed on the need to 'see' and 'understand' systems (see Table 6.8). In other words, there was a sense among some that the resolution or failure of larger problems is contingent upon people's ability to see things from a systemic perspective.

Another categorical way that system was used resembles the system in the Triple–S view of the pest management discussion (see Table 6.9). That is, it is the ominous entity that is stultified for its heartless control of present practices. On the other hand, systems are not necessarily a dominating force. Instead, they can be manipulated, crafted, changed or omitted (see Table 6.10). Finally, Table 6.11 depicts people's understanding of general systems theory with its fundamental notion of a group of interacting parts that form a unified whole.

A question of scale

The stakeholders of science were very familiar with systems notions, particularly biological and ecological systems. Sustenance stakeholders were the stalwart defenders of systems thinking from a more inclusive scale. While the scientists freely discussed the ecosystems immediately related to crops and livestock, the sustenance stakeholders spoke of systems on a global scale – in particular, the present and future well-being of people as cited by an organic farmer:

> *We don't really, in talking about prevention, we don't look at the system in a more complete way and that is what I want to see done more of. Look at all the pieces of the system and I guess if I had to*

Table 6.5 *A wide range of systems were identified*

Agricultural system	Food system	Moving system	Stable system
Alternative systems	Human system	Natural system	Ration systems
Biological systems	Just in time delivery system	People system	University systems
California system	Management system	Pest-control system	Water systems
Economic system	Mental health systems	Rating system	Weed systems
Educational system	Money system	Rotational-grazing system	Well systems
Farming system	Monocultural systems	School system	

pick one concern I have about current agriculture, the design of our current agriculture system, in many ways it leaves people out of the equation. That's why you see abandoned farmsteads and rural communities not being as viable as they once were because they are not a part. It looks to me as if the social aspect of agriculture is not considered as part of our design systems.

Sustenance frequently expressed, with great exasperation, the lack of systemic thinking evidenced by short-sighted decision-makers. In doing so, they became selective of who and what is considered to be part of the system. They neglected the fact that there are people in the dialogue with competing needs and desires (for instance, manufacturing, banking, government). These people cannot be dismissed from the system once identified as the enemy of their cause.

Clearly, the notion of systems comes as second nature to sustenance stakeholders due to their strong ecological proclivities. Sustenance is obviously not against food production; however, they desire production to be done with as little disturbance to the balance of nature as possible. The dominant practice of agriculture represents a massive threat to that balance. To make matters worse, when the primary motivation of the invasion is financial gain, a system advocate is customarily seen as a covetous foe. Obviously, this adversarial dismissal is not limited to sustenance stakeholders.

Note to innovators

When it comes to the development of a sustainable innovation that considers all stakeholders, it is imperative that the innovator gains a bird's eye perspective from the highest level possible. This large-scale perspective allows the innovator to see the breadth of the system that includes the economic concerns of system, utilizes the discoveries of science, and limits the threat to nature's balance as described by sustenance.

The innovator must not ignore anything but must also avoid being seduced by stakeholder eloquence. If the innovator draws value judgements on which needs and desires are more relevant or noble than another, the ignored need will emerge at a later and more costly stage when developing the innovation. Once the needs and desires have been identified, defined and understood in terms of importance, the innovator is in a better position to satisfy not only the needs of the customer, but also those of stakeholders related to the problem.

Table 6.6 *Condition of systems*

Beat-up systems	Diverse systems
Broad system	Entire system
Closed system	Imperfect system
Complete system	Status quo system
Complex system	Unstable system
Different systems	Large system

A question of the head or the heart: Education as solution?

Wisdom is ever a blessing; education is sometimes a curse.

John A Shedd, *A Salt From Any Attic*, p29

There is an ironic similarity of stakeholder opinion on how to solve problems. For many of the stakeholders, the sorry problem that sustains the tension in the pest management dialogue is that people outside of the stakeholder's circle simply lack a 'complete' understanding of the issue. Consequently, the obvious solution to the pest management dilemma proposed by most of the stakeholders is simply to educate their opponents.

Organic farmer — *And there has to be some way to make larger sense out of things. The only way I know how to do it is to educate people so that they understand systems and they understand relationships and we've only just begun to do that. Until we're successful at doing that, we're not going to be able to make better sense of what goes on around us.*

But who should be educated? The following quotation that supports the need for education is from a conservationist and an environmentalist:

Conservationist — *So I think we need to give, an education has to be in the legislature, has to be with the farmer, the bankers.*

Environmentalist — *I get tired of always hearing about educating farmers, educating farmers. It always falls on their back to do this but there are a lot of other people that need to get into this education component. The user, the consumer and, I think, quite frankly, the corporate forces, the financial forces...*

While trying to support the organic market, a produce purchaser for a grocery chain stated that it is difficult to be successful in this endeavour when devoid of education:

Retail — *But the consumers, yeah they all talk it, they want organic produce, they want stuff that's safe to eat, but right now, and I feel it's due to a lack of education, they are not willing to pay the price to get it. Until we get more consumers that are willing to do that, and support it, it's going to have a hard time getting off its feet and get going.*

The call for education does not only come from sustenance stakeholders. An entomologist agreed that education would lead to greater tolerance of pests:

Table 6.7 *Active systems*

- System which protected the health of the consumers...
- Chemicals are more of a correction to that system...
- The system that is alive out there somewhere...

> *And getting back to education, for students to understand this, if people understand this a little bit more, that they do have a place, an ecological place, then we can tend to tolerate [pests] a little more.*

The system also accepted that education could lessen the fear occasionally experienced by the general population. One way to reduce the fear, as reported by a scientist who works for a chemical manufacturing company, is to inform people of the vast amount of care that goes into the testing of the various chemicals:

> *And as your concern is, am I getting all kinds of residues on my fruit, we're doing thousands and thousands in hours and years of testing before that piece of fruit is treated with that pesticide... We haven't communicated to [the urban-dwelling consumer] that that food is safe. It is more dangerous for her to put gasoline in her car than to eat 12 tonnes of apples treated with Alar. But that never got across.*

A corporate executive in charge of planning for a major chemical manufacturing company believes that people need to be informed on the costs and benefits of chemical usage:

> *And I think [executive of environmental public affairs], you said it well earlier, it's education. If people understood people, then they'd see that it isn't good or bad, it's degrees. I mean you gotta say an environmentalist has more concern about nothing but the environment than we do because we see it differently than they do. We clearly do. We see that, yeah, every now and then we can't lie and say that a spill in an ocean is not going to have some detrimental effect on fish. It is. But to never have that spill in there costs you so much on the other end in terms of what [executive of environmental public affairs] said, you've got to hand hoe weeds, your production level goes down to about half of what it is today. Half to feed, the food costs will triple. So there's so many benefits on the other end that you take the risk of having that possibility of some negative effect of one event on the environment. The environmentalists don't see it that way at all.*

This executive further applied the issue of education to the triple–S of the pest management dialogue:

> *Those people [that is, science] have different interests in learning and obviously have different motivations. The last group [that is, sustenance] is not so interested in learning as they are in*

Table 6.8 *Systems in cognitions*

Belief systems
Consensus on systems
Popular systems
Systems perspective
Systems talk
Systems thinking
Understand systems

teaching. The second group [science] is probably more interested in learning because it's in their best interests but that's to pacify both groups on both sides. It's difficult as [the executive of environmental public affairs] said, I mean, how much money and time do we spend in trying to educate people. You know, our interest is their interest. We don't have any interest in harming anyone.

There exists a consensus among most stakeholders that education is the obvious response to their problem. However, there also exists a sense of acquiescence among the stakeholders that this generation cannot be reached, in spite of the best attempts. The systems therapist inquired of the corporate executives why they felt education does not work:

Therapist	*So what prevents resolution of this problem? What's preventing people from hearing your story?*
Public	*Not enough money to get the story out there.*
Planning	*Plus, as you said before, people have to want to hear it...*
R&D	*That's right.*
Public	*Yes.*
Planning	*...They have to want to listen and be attentive. It's an on and off thing. You know, all of a sudden when you hear Alar hit the newspapers, your level goes up. And then you have a long pause before you're interested again.*
Public	*You lose a few years of time.*
Planning	*How many people are involved in agriculture today in the whole US? So it's not like it was in the fifties and forties and thirties where predominantly it was people-related, people could relate to it.*
R&D	*It's like the headlines. You can picture atrazine found in 72 per cent of the wells in the midwest. Now, for us, we can't use that same number of words to get the truth back up there. And that's what we were talking about earlier. You've got to just generate pounds of information to get to that. And some people like [the executive of environmental public affairs] are good at that, if you get the right person to listen to that. So what is the significance of that?*
Therapist	*When you do have a captive audience, you present the truth to them and they've got it in their hand. What tends to happen after that?*
Public	*Well, if it's an audience of ag broadcasters, some of them deliver the message out that they gathered from that session. If it's grade school kids, you hope they retain some of that as they go through life. If it's a group of California Department of Food and Agriculture government people, maybe they'll incorporate that into their daily life. Farmers, I think, hang on to some of that. It just depends on the audience and what they do. If it's just people off the street, they probably don't remember very long. I dunno.*

In the last statement, 'public' suggested that school kids should be educated. This is consistent with other stakeholder views that the next generation should be educated at a young age. One of the organic farmers said:

The fundamental problem seems to be, then, that we don't make people think in the long term. The educational systems don't work that way. If we need to change something, it's that. If you want to address the problem at a fundamental level, that's where I would address it. We have to teach people, teach agricultural students that there is more to agriculture than maximum yields. The concept is teaching in whole systems.

Table 6.9 *An organized society or situation frequently regarded as 'the system'*

...but the system you are talking about is supported by the legislature and Iowa state research...
...buy into the system...
...have learned how to play the system...
...hooked into the system...
...I don't have a lot of trust or confidence in the system...
...I don't think that system is sustainable any more than the communist system is sustainable...
...is the system supportive of that...
...know how to work the system...
...money runs the system...
...people in the system really don't give a damn...
...the system is set up so that we don't talk about the common interests...
...trusting the system...
...we're dependent upon the system that has built-in uncertainty and insecurity...

Additionally, the produce purchaser for a grocery store chain put it this way:

> *I think it all comes back to education. We have to do more starting in the classrooms, even in junior high and high school, in educating the young kids going into adulthood on what to expect, how to organically grow better produce, and let 'em realize you are not going to get the yields off an organic field that you would off one which uses chemicals. Which goes back to how can you afford to make a living when you're only getting ten bushels an acre yield instead of 40 bushels an acre.*

He supported his claim with the following:

> *But, as a whole, in a retail outlet ... in one of our combination stores or bigger stores...the biggest towns it goes over in are college towns...because I feel those kids are educated and know the difference and are willing to pay the difference to support a programme like that. I think that when you get away from that, the education isn't there anymore. For the most part, no, it doesn't have the support of the consumers.*

The chemical manufacturing executives of environmental public affairs and research and development for crop protection stated:

Public	*The industry keeps trying to come up with some things. You know they've got the Benny Broccoli story now that they're putting out in the school systems. Just trying to educate kids a little bit about agriculture and gardening and pests.*
R&D	*Right.*

Upon hearing the discussion of education, the sociologist from the reflecting team sceptically wondered if education, instead, is an attempt to recruit people into the viewpoint of the teacher:

> *Yeah, or it's education as I see the world... What is education and what's the content of that and what's the consequences of that? We have some endpoint in mind and the endpoint is kind of perpetuating the view that I have.*

Table 6.10 *Systems are dynamic*

Build their complete system on trust
Change a system
Create the system
Design a system
Design our systems
Didn't have these chemicals in our system
Disruption in the system
Grapple with systemic problems
I've extracted myself from the system
Inputs into that system
Jump out of the system
Modifying a natural system
Overcome the system
Pushed that system
Redesign a system
Restructure the system
Something else has driven the system
Tailor-made to that system

This comment by the sociologist was fed back to the executives of the chemical manufacturing company. The executive of environmental public affairs said the following:

Public	*...So it comes down to education. And you sorted out education as either education or, what was the other word? You're either trying to educate 'em...*
Therapist	*Recruitment?*
Public	*...or recruiting. That's what education is, it's recruiting. I wouldn't say that one excludes the other there.*

It should be noted that the media were frequently indicted as failing to report the 'true' story. How did members of the media see their role in the pest management debate? One member of the print media that participated in the stakeholder group said:

An important thing for the media is to be useful in terms of educating as many people as possible...That's, I think, we should be expected to do that – educate.

Education was the best remedy the stakeholders could produce to reduce the confusion and misinformation regarding pest management; however, few were optimistic of this option. Again, the sociologist from the reflecting team observed that perhaps the problem is not merely the result of confused cognitions, rather, it is the result of a self-serving culture:

...And there were a lot of things there that at one point I thought we were having a discussion on the reflections of a hedonistic society. And because I think in many ways that is what our society is, and therefore it makes it really difficult to change course and do things when the interest is on the self rather than on the collectivity. And I see that more and more and not just in food, in the food realm and agriculture realm, but in all aspects of our culture.

Table 6.11 *The whole and its parts*

Checks and balances in the system
Does not look at the entire system
Familiar with the rest of the system
Focus on our piece of the system
Other parts of the system
People's various roles in the system
Pieces of this system
Putting it back into the system
Sows always nurse their young and that's part of the system
The system isn't an integrated system
There were all kinds of interreactions and reactions between systems
We don't look at the system in a more complete way
What's going on within a system
Whole systems

An urban-dwelling consumer echoed the same note made by the reflective sociologist:

Consumer — *And whatever drives convenience, whether it be greed or fast pace of life or whatever it is, I'm not sure where all that comes from.*

After some discussion, the participants concluded that money is what drives convenience. Subsequently, the curiosity of the environmentalist created a pivotal point in the group's dialogue of the group:

Environmentalist — *The question is, why do people live like that?*

Therapist — *Live like what?*

Environmentalist — *What is the self-gratification, the good life? Why is it necessary that we have the good life? The immediate self-gratification of everything that we want all rolled immediately into it.*

Consumer — *In a word, I'd say self. And I'd say this comes down to the human condition, which is kind of a spiritual thing; the premise that man is basically good. I don't necessarily buy that. I think we have good desires and things, but there is a thing called self that tends to get exalted above all things. We tend to do that. And it's not until you become selfless or start to see a long-term investment in the kind of the good of the whole versus the good of yourself. You're not going to be willing to do that. And I think our society over the past years, especially over the eighties decade, in general, was a very selfish society. And I think we're starting to see the benefits of what that's got us; a bunch of junk. We're reaping what we've sowed. Now we've got a lot of things that have been set in motion. I believe that farmers are probably typical and from a historical standpoint, probably some of the most conscientious people, some of the most patient people, and really care about being good stewards of the land, literally. But we've evolved into a society where technology has taken over and convenience is pushed. And they're getting pushed faster and faster. And yet, everyone wants a piece of the American dream. Everyone wants a good life. In order to survive, you've got to go to some of these things. Now we're at a crossroads.*

Interestingly, no one contradicted or challenged the consumer's chilling comment. The obvious lack of response or resolution to the 'selfish human condition' remained ever-present throughout the rest of the second and third sessions.

Conclusions

If input towards innovation is only gained from the customer, the likelihood of unforeseen problems is greatly enhanced. Some innovators have recognized this but opt to avoid the hassle of contending with stakeholders. The delivery of a sustainable innovation requires input from all the stakeholders who are involved in the issue, although this is no easy task. Stakeholder input is acquired in numerous ways, typically through the use of focus groups. In this chapter we have attempted to highlight some of the problems frequently incurred by using traditional focus-group methods. In response to the multiple problems, we have introduced systems therapy, a method for understanding and creatively thinking about the complexity of the social system inherent to the pesticide discussion. We have been developing this rigorous process to provide utility while maintaining integrity, accountability and cost-effectiveness to stakeholders. We have attempted, furthermore, to take a step towards addressing issues relevant to the pest management dialogue. The primary goal of this study was to gain an understanding of the human interaction that occurs when stakeholders of pest management interact. A secondary goal of the study was a search for sustainable business opportunities for pest management although it was never expected to be fully achieved. Three 90-minute sessions with a heterogeneous bunch of stakeholders is certainly not enough time to fully understand their needs and to design a product. The process is not complete and requires further iterations between stakeholders and designers. On the other hand, the study gleaned an improved understanding of this difficult issue that is extremely useful for planning the next iteration.

The Triple–S description of system, science and sustenance provides an improved and more accurate view of the pest management dialogue than does the dualized description of 'us versus them'. We saw that each of the three parts operates from vastly different paradigms, motivations and incentives. A coalition between system and science tends to limit sustenance from participating in the development of new technologies and, therefore, new paradigms of farming practices. System and science should further cultivate relationships – with assistance from science when necessary – that could lead to a greater satisfaction of all stakeholders.

Although systems talk is currently in vogue, people tend to set their boundaries around diverse problem spaces and on different scales while accusing others of not being systemic. This creates communication problems. As a result, another stakeholder's ideas, if dissimilar, are in peril of being foreclosed upon. When designing a sustainable innovation, we are not convinced that all stakeholders in a system need to fully understand the macrosystem. As long as each stakeholder's needs, desires and paradigms are truly considered by the innovator, the larger system will have a greater chance of being represented.

Has education proven to be the universal remedy for all that befalls mankind? Education provides us with information on what to think and how to think, but it has its limitations on how we *choose* to behave. It is at the moral crossroads that we discover that we have been grappling with problems that have been poorly posed. It must be remembered that providing food and fibre in a sustainable way is a difficult problem, replete with manifest paradoxes: development or harm, personal need or care for others. We are reminded of the axiom that unless a grain of wheat falls to the ground and dies, it cannot bring forth fruit. What do the stakeholders bring to the discussion table. What fruit has it borne?

The discussion of pest management is very volatile with such high stakes content as food, water and soil. It is possible for society to demonstrate principled reasoning and consequent consensus, but the political underbelly of mankind becomes grossly apparent when it comes down to practice. When policy creation is attempted or decisions of practice are required, truth is often confused with preference. Pride, covetousness, constricted thinking and one's intoxication with personal agenda all confound any progress towards the common good.

Given this situation, an approach is required that can transcend even social movements and common causes – an approach that recognizes the inherent human complexity. We are confident that the pesticide industry is ripe for developing sustainable products, but this can only be achieved if it dares to find a way that is responsive to the multiple stakeholders. The achievement is contingent on the ability to incorporate the often ignored and ever-present source of all human complexity – *personal values.*

References

Alessi R and Mayhew M (1995) 'Teaming techniques derived from human systems theory: Neutrality, hypothesizing and circularity' in Schoening B and Wittig B (eds) *Systems Engineering in the Global Market Place, Fifth Annual International Symposium Proceedings of the National Council on Systems Engineering (NCOSE)*, July 23. St Louis, Missouri.

Anderson H and Goolishian H (1988) 'Human systems as linguistic systems: Evolving ideas about the implications for theory and practice'. *Family Process*, 27, pp 371–93.

Badaracco J (1991) *The Knowledge Link: How Firms Compete Through Strategic Alliances.* Harvard Business School Press, Boston.

Badaracco J and Ellsworth R (1991) *Leadership and the Quest for Integrity.* Harvard Business School Press, Boston.

Bakhtin M (1981) *The Dialogic Imagination.* University of Texas Press, Austin.

Bakhtin M (1986) *Speech Genres And Other Late Essays*, University of Texas Press, Austin.

Baltes P, Reese H and Lipsit, L (1980) 'Life-span developmental psychology'. *Annual Review of Psychology*, 31, pp 61–110.

Bateson G (1971) 'The cybernetics of self: A theory of alcoholism'. *Psychiatry*, 34, pp 1–18.

Bateson G (1972) *Steps to an Ecology of Mind.* Ballantine Books, New York.

Bateson G (1979) *Mind and Nature: A Necessary Unity.* EP Dutton, New York.

Bateson G, Jackson D, Haley J and Weakland J (1956) 'Toward a theory of schizophrenia'. *Behavioral Science*, 1, pp 256–264.

Becvar D and Becvar R (1988) *Family Therapy: A Systemic Integration.* Allyn and Bacon, Boston.

Bernard H (1994) *Research Methods in Anthropology: Qualitative and Quantitative Approaches.* Sage, Thousand Oaks.

Bertalanffy L (1968) *General System Theory: Foundations, Development, Applications.* George Braziller, New York.

Bowen M (1976) 'Theory in the practice of psychotherapy' in P Guerin (ed) *Family Therapy: Theory And Practice.* Gardner Press, New York.

Braten S (1987) 'Paradigms of autonomy: Dialogical or monological?' in G Teubner (ed) *Autopoiesis in Law and Society.* EUI Publications, New York.

Broadhead R (1983) *Private Lives and Professional Identity of Medical Students.* Transaction, New Brunswick.

Bruner, J (1990) *Acts of Meaning.* Harvard University Press, Cambridge.

Cronon, W (1992) 'A place for stories: Nature, history, and narrative'. *Journal of American History,* 78, pp 1347–1376.

Davies, P (1992) *The Mind of God: The Scientific Basis for a Rational World.* Simon & Schuster, New York.

Denzin N and Lincoln Y (1994) *Handbook of Qualitative Research.* Sage, Newbury Park.

Erikson E (1963) *Childhood and Society* (2nd ed). Norton, New York.

Erikson, E (1964) *Insight and Responsibility.* Norton, New York.

Flood R and Jackson M (1991) *Creative Problem Solving: Total Systems Intervention.* Wiley & Sons, London.

Geertz C (1973) 'Thick description: Toward an interpretive theory of culture' in C Geertz (ed) *The Interpretation of Cultures.* Basic Books, New York.

Glaser B and Strauss A (1967) *The Discovery of Grounded Theory: Strategies for Qualitative Research.* Adline, Chicago.

Glaser B and Strauss A (1970) *Status Passages.* Adline, Chicago.

Guba E and Lincoln Y (1989) *Fourth Generation Evaluation.* Sage, Newbury Park.

Haley J (1963) *Strategies of Psychotherapy.* Grune & Stratton, New York.

Husserl E (1970) *Logical Investigations.* Humanities, New York.

Kohlberg L (1969) 'Stage and sequence: A cognitive-developmental approach to socialization'. in DA Goslin (ed) *Handbook of Socialization Theory and Research.* Rand McNally, Chicago.

Kohlberg L (1973) 'Continuities in childhood and adult moral development revisited' in P Baltes & KW Schaie (eds) *Life-span Developmental Psychology: Personality and Socialization.* Academic, New York.

Krippendorff J (1980) *Content Analysis: An Introduction to its Methodology.* Sage, Beverly Hills.

Lederer W and Jackson D (1968) *Mirages of Marriage.* Norton, New York.

Lincoln Y and Guba E (1985) *Naturalistic Inquiry.* Sage, Newbury Park.

Lofland J (1971) *Analyzing Social Settings: A Guide to Qualitative Observation and Analysis.* Wadsworth, Belmont.

Maturana H (1978) 'Biology of language: The epistemology of reality' in GA Miller & E Lenneberg (eds) *Psychology and Biology of Language and Thought.* Academic Press, New York.

Maturana H and Varela F (1987) *The Tree of Knowledge: The Biological Roots of Understanding.* New Science Library, Boston.

Mayhew M and Alessi R (1993) 'Responsive Constructivist Requirements Engineering: A Paradigm' in McAuley J and McCumber W (eds) *Systems Engineering in the Workplace* Proc. Third Annual International Symposium, National Council on Systems Engineering (NCOSE), Arlington, VA, 26–29 July, Washington, DC.

Mayhew M and Alessi R (1995) 'Foundations of a human systems theory and systems engineering' in Schoening B and Wittig B (eds) *Systems Engineering in the Global Market Place* Proc. Fifth Annual International Symposium of The National Council On Systems Engineering (NCOSE), 23–26 July, St. Louis, Missouri.

Miles M and Huberman A (1994) *Qualitative Data Analysis.* Sage, Newbury Park.

Minuchin S (1974) *Families and Family Therapy.* Harvard University Press, Cambridge.

Morgan D (1993) *Successful Focus Groups: Advancing the State of the Art.* Sage, Newbury Park.

Morgan, D (1988) *Focus Groups as Qualitative Research.* Sage, Newbury Park.

Namenwirth J and Weber R (1987) *Dynamics of Culture.* Allen & Unwin, Winchester.

Orwoll L and Perlmutter M (1990) 'The study of wise persons: Integrating a personality perspective' in Sternberg RJ (ed) *Wisdom: Its Nature, Origins, And Development.* Cambridge University Press, New York.

Piaget J (1952) *The Origins of Intelligence in Children.* International Universities Press, New York.

Piaget J (1955) *The Child's Construction of Reality.* Routledge, London.

Radin P (1966, orig. 1933) *The Method and Theory of Ethnology.* Basic Books, New York.

Random House Dictionary of the English Language, Unabridged Edition (1984) Stein J (ed) revised. Random House, New York.

Riessman C (1993) *Narrative Analysis.* Sage, Newbury Park.

Satir V (1964) *Conjoint Family Therapy.* Science and Behavior Books, Palo Alto.

Schafer R (1992) *Retelling a Life: Narration and Dialogue in Psychoanalysis.* Basic Books, New York.

SEI (1993) 'Software Requirements Engineering'. *Bridge,* 2, pp 17–21, Systems Engineering Institute, Carnegie Mellon Univ, Pittsburgh.

Shenhar A (1991) 'Developing the Discipline of Systems Engineering'. *Proc. First Ann. Int. Sym. of the National Council On Systems Engineering* (NCOSE/ASEM), pp 148–149.

Spradley J (1979) *The Ethnographic Interview.* Holt, Rinehart & Winston, New York.

Stone P, Dunphy D, Smith M and Ogilvie D (1966) *The General Inquirer: A Computer Approach to Content Analysis.* MIT Press, Cambridge.

Strauss A (1987) *Qualitative Analysis for Social Scientists.* Cambridge University Press, New York.

Strauss A (1991) *Creating Sociological Awareness.* Transaction, New Brunswick.

Strauss A and Corbin J (1990) *Basics of Qualitative Research: Grounded Theory Procedures and Techniques.* Sage, Newbury Park.

Veatch H (1969) *Two Logics: The Conflict Between Classical and Neoanalytic Philosophy.* Northwestern University Press, Evanston.

Watzlawick P, Beavin J and Jackson D (1967) *Pragmatics of Human Communication.* Norton, New York.

Weber R (1990) *Basic Content Analysis.* Sage, Newbury Park.

Weiner C (1981) *The Politics of Alcoholism.* Transaction, New Brunswick.

Weiner N (1948) 'Cybernetics' *Scientific American,* 179, pp 14–18.

Woods T (1993) 'First Principles: Systems and Their Analysis' in McAuley J and McCumber W (eds) *Systems Engineering in the Workplace.* Proc. Third Annual International Symposium, National Council on Systems Engineering (NCOSE), Arlington, VA, 26–29 July, Washington, DC.

Wymore W (1993) *Model-based Systems Engineering: A Text.* CRC Press, Boca Raton.

7

FUTURE SEARCHING FOR NEW OPPORTUNITIES INVOLVING THE PESTICIDE INDUSTRY AND SUSTAINABLE AGRICULTURE

David Morley and Beth Franklin

INTRODUCTION

Issues such as the future of the pesticide industry and the way in which the concept and practice of sustainable agriculture can affect that future often appear remote and technocratic. Such issues seem to be dominated by inaccessible business, government and scientific bureaucracies, even to those who are directly affected. Partly, this is the result of the lack of forums for debate that involve such stakeholders in a collective consideration of their own futures.

This chapter presents the story of such a forum, with particular emphasis on the content and directions of discussion. It records the outcomes of a particular style of participatory action research as it was used to generate exchange at the complex and often tense boundary between pesticide producers and proponents of sustainable agriculture. The means selected in this project to bring these diverse groups together is known as *future searching*. It is designed specifically to provide settings in which the representatives of the various interests involved in complex and changing issues are brought together to work towards the definition of shared perspectives of the future. The distinctions between pesticide producers and users and proponents of sustainable agriculture reflect this situation precisely. In bringing together the array of interests most directly concerned with alternative agriculture *and* those concerned with the research, manufacture, and marketing of chemical pesticides, the future search process engaged opposing points of view representing one of the most significant fractures in late 20th-century society. This is the fracture between the positions of an established modern industrial complex and alternative environmental perspective of a 'postmodern' future.

The essence of the competing issues relates to a different association with the environment; each perspective of the term 'environment' appears to be in direct opposition. The pesticide industry is concerned with environment-as-resource, and its actions are aimed at utilizing that resource to the maximum benefit of producers and consumers who depend upon it. The concept of sustainable agriculture is based on the notion of environment-as-ecosystem, the preservation of which is dependent on a long-term maintenance of its natural capacities of regeneration.

The Options for Pesticide Producers and Regulators in Sustainable Agriculture project, based at the Leopold Center for Sustainable Agriculture at Iowa State University, recognizes that society is rarely provided with the institutional contexts for effective debate and collaboration around boundary issues. The collision of values and beliefs that occurs across the societal divide is mostly loud and confrontational. More accurately, there is rarely an opportunity for an open, collaborative expression of views. One broad conclusion of the project has been the recognition of the urgent need to change the relations between factions. In summary, the project maintains that this relationship needs to be:

- proactive rather than reactive;
- holistic rather than specialized;
- collaborative rather than competitive;
- carried out in more effective public forums.

These directions for shared process have been proposed because we believe that the futures of both the pesticide industry and sustainable agriculture are inextricably linked. The purpose of the future search conference, which is the focus of this chapter, was to explore the nature of this connection as it is understood by the people who are involved in shaping it.

FUTURE SEARCHING

The principles and objectives that lie behind the idea of future searching provide important guidelines for the direction taken by participants and undoubtedly affect the conclusions that may be drawn from the process. Future searches, or search conferences as they were originally known, were invented in 1965 by Eric Trist and Fred Emery, researchers then based at the Tavistock Institute for Human Relations in London. Search conferences, which were developed as a form of active adaptive planning and as an expression of action research, are based on the idea of facilitating planned organizational change through the application of the theory of collaborative relationships (Emery and Trist, 1973, p104). It is argued that through collaboration between outside facilitators (not motivated by personal agendas associated with the problem focus) and the various classes of inside stakeholders, joint insights into the future can be gained. This contrasts significantly with traditional planning for long-term change in which the task is typically controlled by senior managers, delegated to strategic planning divisions, or addressed by interdepartmental committees. In such settings there is rarely an opportunity for all the different interests to participate and it is difficult, if not impossible, to deal with fundamental shifts in settings where the 'rules of the game are changing'; the tendency is to stick to likely or probable futures (Emery, 1993, p231).

In 'searching', the emphasis is on possible futures – even on improbable futures.

Future searching is necessary because 'existing bodies of data and current notions of what is relevant can be no substitute for people's own sense of what is coming over the horizon ... today's probabilities are not a sure guide to the future' (Emery, 1993, p231). Search conferences provide key stakeholders with an opportunity to cross traditional boundaries (time, geography, class, race, ideology, disciplines, institutions) and, thereby, to examine new perspectives in new ways – inventing new futures. Search conferences have been used in many parts of the world to provide new ways of addressing a wide variety of complex and intractable issues, and there is a growing literature which demonstrates the widening application of this approach (Morley and Trist, 1993; Weisbord, 1992; Weisbord and Janoff, 1995; Wright and Morley, 1989). In particular, there has been a significant increase in the use of 'future searching' (Weisbord's term that is now widely used) to address metaproblems – complex, multifaceted problems encompassing many interrelated issues that extend beyond existing organizational and jurisdictional boundaries (Franklin and Morley, 1992; Morley and Trist, 1992). Out of this work has come the notion of a 'contextural orientation' to searching which promotes the integration of all interests associated with a metaproblem in a context that 'favours no group or position over any other' (Franklin and Morley, 1992, p231).

As a facilitated process, future searching is based on a series of general process principles:

- It is an open participatory process which aims to involve participants.
- It emphasizes equal access to the process for all participants and encourages people to listen to each other and to work together in defining a common ground that allows the group to move the process forward together.
- It draws strength from the multi-interest character of the domain that is represented by the differences among participants.
- It encourages participants to work at the boundaries of what is known and experienced and what is unknown and uncertain, stable and unstable, ordered and chaotic.
- It draws on the collective commitment, visions and hopes for the future that motivate the participants in their homes, communities, networks and organizations.
- It encourages the rethinking of the standard roles and relationships implied by the participants' institutional and community origins.
- It encourages a sense of ongoing learning among participants and increasing self-confidence in their collective capacity to establish directions for future change, to define options that are viable, and to initiate continuing self-organization among the institutions and interests present.

Future search events are designed to provide settings for open debate in which invited participants are encouraged to express their own views on the issues under consideration, to listen to others approaching the issues from alternative perspectives, and to explore the existence of common ground among the different views in order to collectively define desired futures and ways to achieve them. 'Searching' is just that: a search for alternative futures that involves breaking new ground in areas that we have not previously experienced. Strong differences in opinion are not denied or avoided in future searches. This is not a form of traditional conflict resolution or arbitration where compromise and the 'working out' of differences are the objectives.

Futures searches are also social events. They maximize the mixing of interests in small group settings where everyone has extended air time; they bring participants together in plenary sessions in which the work of the small groups is exchanged and compared; they provide spaces for continuing informal discussion of the emerging collaborative themes and for relaxation and fun. They form a temporary 'social island' – especially when taking place over two to three days in a remote residential site. But the intention is always to encourage the design and planning of future collaborative actions that can extend the results of the search beyond the actual event.

FUTURE SEARCHING THE PESTICIDE INDUSTRY AND SUSTAINABILITY QUESTION

The Options for Pesticide Producers and Regulators in Sustainable Agriculture project provided the context for the search event that produced the material discussed in this chapter. Thirty-six people were invited to gather at the Koinonia Retreat Center, South Haven, Minnesota, during June, 1996. They represented the pesticide industry, industrial users of food and fibre, government regulatory bodies, non-profit organizations, scientists, sustainable and organic farming organizations, crop consultants, practising farmers, watershed associations, university-based researchers, and the public.

The participants worked together through the four phases of the future search event in small working groups; the phases comprised:

- an *initial scan* of the general situation that is jointly relevant to the pesticides industry and sustainable agriculture;
- an exploration of the long-range visions and alternative *desired futures* associated with the topic;
- a balancing of *constraints and opportunities* that hinder and encourage the achievement of those goals;
- the generation of key *tasks and actions* that form the basis for continuing activity among the various stakeholders.

In each case, different perspectives generated by the participants were actively discussed in plenary sessions and the potential for consensus and joint action was examined. Figure 7.1 provides a summary of the elements of the future search and their timing. The inverted triangle demonstrates the way in which a search process begins with a holistic, overall review of the issues and encourages the participants to gradually narrow the discussion to manageable task areas for future action that are agreed by the group.

ORGANIZATION OF CHAPTER

This chapter focuses on two dimensions of the outcomes of the Minnesota search: the particular way in which this cluster of participants framed the issues associated with future joint options for pesticide producers and sustainable agriculture practitioners; and the creative strategies developed by the participants with the objective of creating an alternative future for these domains.

The language of the chapter is drawn directly from the discussion in the sequence of sessions at the future search. In this sense it is expressed in the participants' own words, as they were recorded by the facilitators. The orientation of their proposals, the allocation of priorities and the defining of the strategic processes necessary to achieve their objectives are all based on material generated at the conference. This means that it is strongly influenced by the actual group of participants who attended. However, the

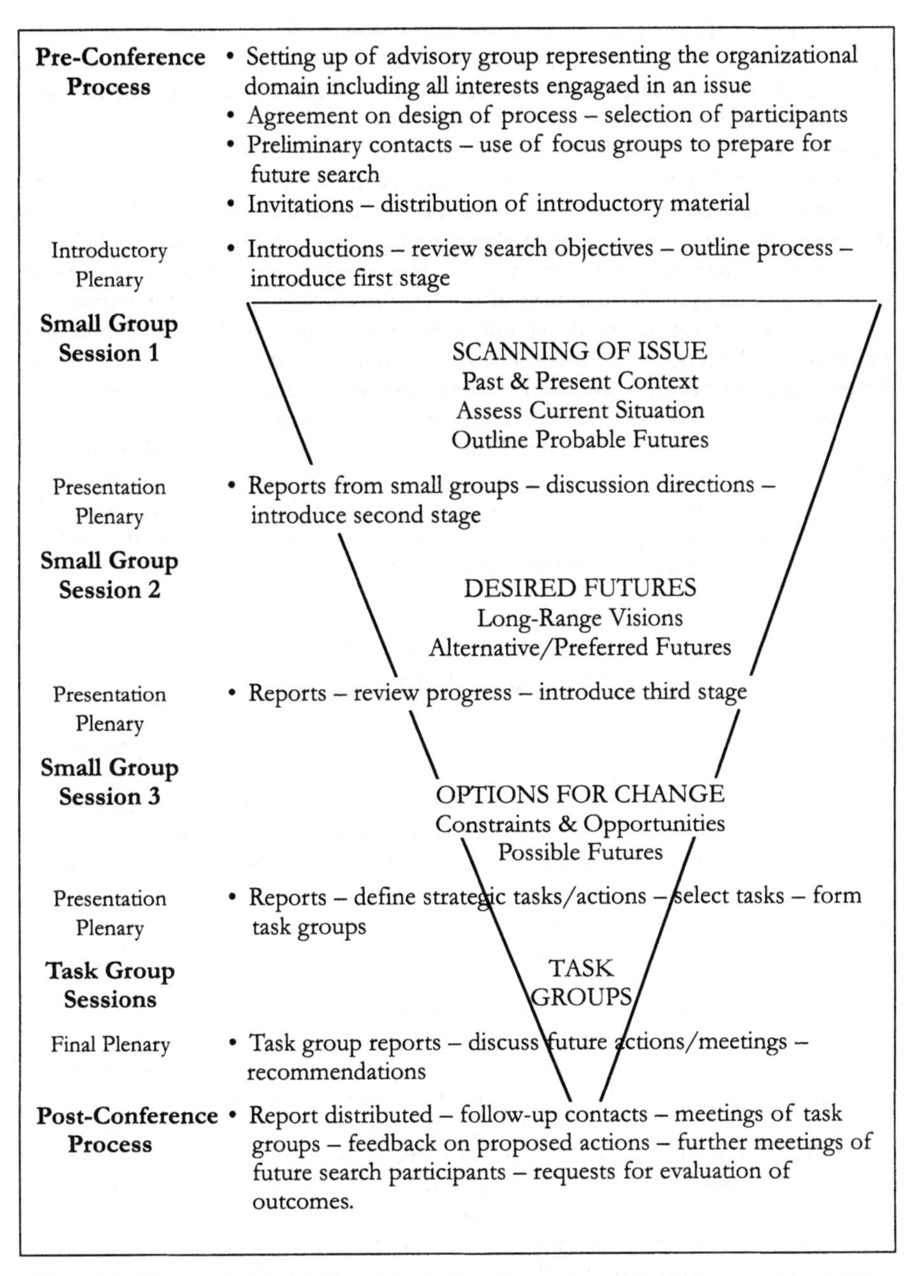

The use of an inverted triangle is intended to reflect the way in which the future search participants begin by establishing the total context of an issue (past and present) from their collective perspective and then work through the search sequence, which allows them to focus on the strategic tasks and actions that can carry the process forward. The search is for sustainable collaborative activities that can continue this adaptive practice through further search cycles.

Source: The ABL Group

Figure 7.1 *Future search design*

organization of the material has been the responsibility of two groups – the small-group facilitators at the event who recorded discussion on flipcharts and the authors who were both group facilitators and overall designers and managers of the process. This organization of the discussion and its outcomes is based on an analysis aimed at reflecting the context and direction of debate.

The chapter begins with a summary of the statements made by participants in the initial scanning stage of the search. This is followed by an examination of key themes that frame discussions in the small working groups throughout the future search. Table 7.2 provides a summary of the nine themes. They evolve from an analysis of current issues to a framing of desired or ideal futures, the consideration of the barriers to achieving these objectives and the counteracting opportunities available for action, and finally to the defining of key tasks. The next section examines the final stage of the future search in which the participants attempt to bring together the strands of discussion around selected tasks that demonstrate the potential for the process being continued in various selected settings. Details of the selection of these task groups are followed by a review of their activities. Finally, the conclusion focuses on the more general implications of the search process for the pesticide industry and sustainable agriculture interests. In particular, it discusses the strong orientation of the participants' conclusions towards a continuing process of searching for alternatives.

Search Themes on the Issue of Pesticide and Sustainable Agriculture

The themes that underlie the continuing dialogue of the future search emerged early in the initial scanning phase. This opening discussion focused directly on the perceived boundary that lies between the two dimensions of the issue under consideration – pesticide producers and users and sustainable agriculture interests. This characteristic of the debate was defined immediately. It was recognized that the contrasting positions spanned the range from those arguing that 'the system is on track' to those asserting that 'the system is in need of fundamental change'. In respect to this wide discontinuity, it was noted that, for this group, the key question was: 'Is there a middle ground between the two positions?'

A further expression of this complex issue was the frequently repeated point that the pesticide and sustainability debate was not just a question of the environmental impacts of pesticides versus biological controls, but also their social and economic implications. Such a broad-based approach to the issue of pesticides and agriculture focuses on quality of life and sustainability as a central theme throughout the future search. This was seen as a perspective comparable to the current concern with 'healthy lifestyles'. This focus emphasised the long-term objective of developing an agricultural system that relies less on chemicals – a position that confronted the values of those who argued that such goals are unrealistic and naive.

This dialectic became clear in the initial scanning stage of the search process when those developing the quality-of-life position had to respond to arguments that focused on the apparent contradiction inherent in the assumption that creating whole-system changes to reduce the environmental impact of agricultural practice involves increasing the agricultural production on fewer acres. Put directly, this asks the question: 'How can we achieve sustainability without cutting yields or increasing the costs of production ... [and] ... at the same time recognize that the operations of large corporations will continue to dominate agriculture in the future?'

Table 7.1 *Creative tensions*

Risk	*Choice*
Economics	Ecology
Consumer demand for cosmetics	Consumer demand for organic products
World food production needs	Risks to human health
Risk perception	Risk reality
Making a living	Producer health
Individual need	Whole-systems need
Urban consumer ignorance	Consumer demand for information
Third World lack of regulation	US regulation (harmonization)
Chemical pesticides	Biopesticides
Smartness	Wisdom
Monoculture (easy to farm)	Biodiversity (more stable)
Global food systems	Local food systems
Focus on system parts	Focus on whole system
Innovation held back	An innovative climate
What's the bottom line?	What's the vision?
Doing same things better	Doing things in a new way
Private choice	Public goals (biodiversity, ecosystem)
Implement new systems	Need to know nature of future problems
Faith in progress, competition, science	Science as a human tool

This overriding challenge to the proponents of sustainable agriculture evoked two primary responses. Firstly, it was recognized that the cheapest (in the short term) and simplest solutions to the problems of maintaining inexpensive food production are still chemical and that the burden of proof for viable alternatives is not on the shoulders of the pesticide companies. This argument is based on the fact that the pesticide industry is highly influential and is constantly proposing the registration of new chemicals and genetics that reinforce the current objectives of the industry. The participants were clear that, being represented by powerful multinational corporations, the pesticide industry was capable of using its lobbying capacity to influence government responses and, therefore, to monopolize the competitive environment. This understanding of the current situation led participants to the conclusion that unless alternative directions for agriculture could be developed that make sense within the pesticide industry, its mainstream practice will not change. A key question facing the group emerged: 'Can a win–win balance be created between sustainable agriculture and agricultural business?' At the same time it was recognized that: 'change cannot come from within agriculture alone'; the backing of the wider society is imperative.

With such a small proportion of the nation's population engaged in agriculture, shifting the paradigm from agribusiness to a sustainable agricultural society will be very difficult. Such fundamental change, participants felt, will have to come from increased public education, involvement and empowerment at the local level. People will have to understand that such a shift demands the application of an alternative economics that incorporates the hidden costs of the current agribusiness model, both to society and the environment. Such an understanding has to recognize that we are all in the same crisis-oriented system which requires the application of a whole-system perspective with long-term dimensions. For one working group, the outcome of its scanning discussion was framed through the recognition of a set of creative tensions which they felt defined the sustainability and pesticides boundary and which will constantly have to be addressed if progress is to be made in creating effective shared arenas of action (see Table 7.1).

Table 7.2 *Future search – key discussion themes*

Scanning Stage	*Desired Futures Stage*	*Constraints & Opportunities Stage*
1 Agricultural perspectives		
• state of agricultural system • family farming • farmers and pesticides	• sustainable farming	• processes of change in agriculture • food security
2 Pesticide industry and the use of pesticides		T3 Encouraging business to be more responsible TASK GROUPS I & II: Stewardship & Business
3 Research, science and technology		TASK GROUP III: Appropriate Technology
4 International issues 5 Role of consumers 6 Education & public awareness		T4 Continuous and lifelong learning
7 Government action and regulations		
	8 Role of local communities • food security/healthy foods • local markets	T5 Strong local communities T6 Bioregional perspective of communities
	9 Creating new institutions and change processes • system-wide dialogue/ trust	T7 Creating a facilitative environment of trust T8 Local ecosystem-based activity TASK GROUP IV: Creating Watershed Management Authorities

Columns indicate the three stages of the future search event.
Titles numbered 1–9 are key discussion themes discussed in the chapter.
Bulleted points are sub-titles used in the chapter.
T3-T8 are titles of organizing tables from constraints and opportunities sessions.
I–IV are titles of task groups used in final stage of the future search.

The future search process is basically a search for common themes of concern that can become frameworks for continuing joint action at the boundaries of change. In this case, themes emerged that reflect the tenuous balance between confrontation and disengagement which inevitably exists between the proponents of sustainable agriculture and those supporting the agricultural status quo. These interlocking themes emphasize the fact that the conflicting dimensions, in fact, are elements of the integrated whole problem. This understanding is reflected in the consistency of the discussion that took place in the four small groups that worked together during the future search.

The following sections deal with the key themes of the search as they were generated and extended from the initial scanning phase, through the expression of desired futures and the assessment of constraints and opportunities, to be refocused in the concluding task group sessions. Table 7.2 summarizes this sequence.

Agricultural perspectives

This area of discussion reflects the participants' assessment of the overall agricultural system as it is viewed from the pesticide and sustainability boundary. Comments on the state of the agricultural system focus on problematic aspects of the situation. Farmers were seen by some to be embedded in a megasystem in which sustainability 'is viewed as antiprogress and antipatriotic'. For such participants, the system's trend is towards increasingly larger farms implementing agribusiness strategies that many see as harmful to the environment. Farmers were viewed as being under considerable pressure to generate the high yields on which their hopes for profitable farming are based. It was suggested that there is a constant fear of slipping into economic loss, together with concern over environmental degradation of the land, declining rural communities and threatened personal health. Behind these fears lies the recognition that individual farmers do not control the business of agriculture – that the future will be dominated by large corporations better able to spread risk. In a world of multinationals running large-scale agribusinesses, it seemed to most participants that the creation of pesticide-free US food systems is still a distant dream.

In the meantime family farming survives, despite the ageing farmer class and the virtual impossibility for new farmers to enter the business, if they are not already connected with farming. Discussion of family farms focused on their importance as agents to develop alternative agricultural strategies. The need to reinforce policies that support all styles and scales of family farming was emphasized. Family farming was seen as a lifestyle choice that is essential for the survival of viable rural communities; it is the basis for sustaining local businesses and services. The diversification option was examined as an important element in the survival of small farms. It was recognized that, even in a period of decline, some new small family farms are being created (for instance, by recent immigrants who locate in niches that agribusiness cannot fill). However, while family farmers thrive on the capacity to be flexible and an ability to diversify in response to a changing market, most feel that they need to rely on pesticides to take advantage of that ability.

The relationship between farmers and pesticides was a recurrent issue for discussion. The degree to which the current generation of farmers is dependent on pesticides was stressed. Farmers fear the risk of chemicals (to health and environment) and want guarantees of safety regarding the products of chemical companies; however, they fear the economic loss that goes with reduced yields even more. Many participants believed that there is a connection between family farming and the development of sustainable farming practices. Many of the commonly recognized issues currently confronting farmers are directly related to the sustainability choice: biodiversity versus toxicity, soil health versus soil erosion, water quality, and the trade-off between sustainable practices and increasing yields. The importance of educating farmers to clarify the choices associated with these issues was emphasized by the participants. It was agreed that farmers want proof that organic farming works before they are prepared to reduce their dependency on chemicals. This demands that sustainable practices must be framed in the context of local ecology, farm economics and social structures, and that farmer education needs to be carried out locally. Participants emphasized that farmers have to be involved in continuous education in order to make effective judgements on what is the best course of action in their particular situation. Such a learning process should be community-wide and should focus on the role of 'agriculturalist as the steward of natural resources'. This stewardship perspective was explored by two of the task groups that were formed in the final stage of the future search; the work of these groups is considered in the concluding stage of this chapter.

An outcome of this broad-based reflection on the current state of agriculture was that participants spent a considerable amount of time on the issue of processes of change in agriculture. Throughout, the duality of the situation was recognized. There are no current indications that the US agricultural system will become pesticide-free as a result of action from within. However, net pesticide-pounds-per-acre levels are down, watershed improvements are gaining ground, and increased public awareness regarding environmental health is having an effect. This provides a rationale for proposals made by this group of participants that large-scale educational, financial and political intervention is necessary to reinforce these changes in the perception and use of pesticides.

Another critical issue which emerged during the future search relates to food security – the maintenance of a stable and healthy food supply system. The achievement of such a condition demands the incorporation of *all* of the costs involved in the economics of food supply (health, environment and social). But this cannot take place through public process alone. There is an urgent need for agrarian reform with these objectives in mind, and the status of the agricultural industry as a whole needs to be addressed at the level of government policy. Emerging public expectations appear to demand an industry which operates in a responsible manner and in which change is related to an environmentally regenerative progress based on systemic and holistic perspectives. This needs to be based on reducing the isolation of farming from the wider urbanized society, a greater focus on resource utilization, and community-wide participation.

The above discussion reflects the overall position taken by the group of participants with regard to the need to apply organic agricultural principles throughout the industry. Participants directly involved in the pesticide industry obviously had some difficulties with aspects of this perspective.

Pesticide industry and the use of pesticides

The participants emphasized that the companies themselves need to learn about alternatives. Reductions in the use of pesticides and the uncertainty of the industry's future often leads to defensive stances being taken and a tendency for decisions to be fear-driven. Nevertheless, it was obvious to most participants that the power of the pesticide industry is continually being reinforced through its ownership of critical information, its control of the directions taken by research and its political clout. A primary conclusion of many participants was that the pesticide industry itself must recognize that the

Table 7.3 *Influencing business to be more responsible*

Constraints	*Opportunities*
• 75% of businesses are publicly owned; stock value drives short-term thinking; share-holders are looking for maximum return. • Most shareholders don't know how their money is being used. • Sustainable techniques and products are regarded as being unprofitable. • Business schools usually do not include this point of view in their curriculum. • Older, established companies find it even harder to make changes. • Majority of profit in chemical companies comes from atrazine and other toxic products – if these are prohibited the cost of production rises significantly – 'dirty' products move wherever regulations are slackest. • The 'silver bullet' mentality prevails – if you have a problem then get a product to solve it. • New product development efforts are often viewed as too risky and may be abandoned. • Businesses do not get exposure to other perspectives, disciplines, values, outlooks, ethics.	• Mutual funds that invest in socially and environmentally responsible business are profitable. • Generate shareholder activism – communicate and inform about ethics. • Organic food processing can be profitable. • Government-established incentives to help ecologically sound companies to get started. • Chemical products are being designed with lifetime of two weeks or less. • New science focuses on interconnect-edness – view the system as a whole. • Encourage development of partnerships to create and support alternative technologies. • New approaches to running businesses more creatively – building a culture that allows innovation. • Personnel from very different organizations generate new ideas, opposing points of view. • Cooperative/joint ventures could offer stability/security to organizations (for instance, cooperation between organic cotton farmers and Patagonia).

system is rapidly changing; the number of farmers using high levels of chemical input are declining and this trend will continue because farmers are increasingly dictated to by food processors who are, themselves, responding to changing, consumer-driven market forces. A fundamental rethinking of the role of science and technology in the reshaping of the future must relate to these changes. However, the participants did not discount the problems inherent in persuading business to adopt the principle of environmental sustainability. One of the small working groups established this issue as a potential field for action during the constraints and opportunities phase of the process (see Table 7.3).

Research, science and technology

Participants recognized that the role of research, science and technology provides a critical bridge between the issue of sustainability and the use of pesticides. From their collective perspective it is, at present, a weak link. Academic research institutions feel threatened by the same power structures that dominate alternative agriculturalists. This was seen to be particularly true of the land-grant universities which are seriously threatened by the loss of grant money. The control of research money and the way it is used was a central issue for this group. It can be summarized in the comment: 'Everybody wants sound science, but since science is bought and sold today, what is credible science?' The sense was that corporate ownership of information in this field has led to it becoming a commodity controlled by very few multinational corporations – particularly knowledge associated with seeds and pesticides. The result was said to be a lack of knowledge regarding alternatives to chemicals; the 'silver bullet' mentality prevails. An important outcome is the lack of consumer demand for alternatives and, of equal importance, the relative lack of connections between researchers, educators and extension personnel, which leads to only limited information reaching farmers.

The role of technology as a reflection of research was seen to be very important. The transformative capacity of chemical technologies and their impacts (economic, agricultural, environmental and societal) needs to be more widely understood. But with this issue, as in most others debated during the future search event, contrasting positions were represented. Accusations of ecological blindness contrasted with the views of those who thought of technology as an end in itself, evidenced in advances in crop breeding and integrated systems of animal productivity. This is examined in more detail in the latter part of the search process by the technology task group.

International issues relating to sustainability and pesticides

Central to the discussion was the fundamental fact that we share the air and water of the planet. The US cannot ignore what is happening in developing countries, especially the distribution of pesticides to the Third World. The lack of regulation, the health implications and the impact of World Bank policies in maintaining the growth of cash crops to pay off national debts were issues discussed as relevant elements of these concerns. The international outcomes of the production and distribution of pesticides were seen as destructive extensions of the American experience regarding pesticide use in agriculture. Global interrelatedness is important, but many of the points made at the search event supported greater autonomy for less-developed countries in deciding what they want to grow, encouraging them to gain access to the basic tools and knowledge necessary to become real players in the future, and especially becoming party to the global impacts of using destructive versus benign chemicals (understanding the trade-offs). A key concluding point regarding the international context, which mirrors positions taken in the US, was the need to support communities and their control of local food systems. This has macro-implications: 'Europe and North America are not going to feed the world.'

The role of consumers

The important role that consumers play in determining the future of food production was raised consistently during the future search. But as in all aspects of these discussions, a number of conflicting perspectives were present: consumers are partly to blame for the continued application of pesticides with their demand for 'the perfect fruit' and access to an unlimited variety of food products; but consumers are also pushing the system to change (for instance, the 'explosion of vegetarianism') based on their growing awareness of the risks of toxicity.

The future search participants do not believe that these contradictions are being resolved. It was suggested that a general feeling of helplessness prevails among consumers; the enormous complexity of the issues makes people feel demoralized, marginalized and fearful. And this fear silences people. So the desired future directions indicated by participants focus on the need to create a civic environment to overcome this powerlessness. The US has many grassroots organizations and an open exchange of ideas is inherent in a democratic society. The question is, how to make this work? Indirectly, consumers are seen to be initiating change regarding organic farming, especially on the West Coast. But it was felt that Europe is far ahead of North America in this regard. An important future direction was seen to be the need for consumers to identify more closely with the farming community. Consumer stakes in the production and consumption of food products are high and they need to develop more awareness of food origins and food safety. Again, this was connected with local food production and encouraging consumers to buy more produce from locally based producers. But this has to be set in the context of a lack of incentives for research into these alternatives, the media's emphasis on crisis situations, and the overall complacency based on the belief that US food is the best in the world.

Education and public awareness

The participants regarded education as essential within all elements of the system: schools, consumers, growers, legislators, and industry. Colleges and universities are important, especially with regard to teacher training, research and the raising of agricultural awareness. It was argued that sustainable practice in the context of food production should be a universal requirement of all educational curricula. Concern was expressed regarding the future of land grant institutions where there has been direct access to decision-making through citizen advisory councils. If land grants disappear it will be the private sector that determines the future of education and research in such institutions.

It was stated that 'learning is central' and particularly the education of children. In one discussion it was emphasized that: 'children need to be informed about how and where their food is produced. This could drive major change by bringing everybody closer to their own food sources. This is the basis for truly democratic choice about technology.' It was suggested that every school should have a garden providing the setting for experiential learning about ideas such as healthy food systems, pest management and sustainability (see Table 7.4). There is a general need to create opportunities to 'learn on the ground' at the local level. Such activity should include open experimentation with alternative approaches to farming that are accessible to everyone.

Once again, constraints were identified in this debate. There is frequently a lack of understanding among parents as to the importance of this aspect of their children's education. This leads to a lack of consensus regarding curriculum change, maintaining the control of existing mandated material, and the limitations to the allocation of resources. Also, it was felt that the shadow of the urban–rural rift in American life is always present in discussions on these issues. Despite these constraints, future search participants returned to the issue of education throughout the process. Educating Americans about sustainable agriculture was essential. Groups concerned with sustainability need to connect with the wider education system (local school boards, state and national institutions) to lobby for the introduction of issues relating to food production, environmental and community sustainability, and bioregionalism into curricula in all educational settings. Similarly, participants indicated that promoting awareness among the general public is critical to the success of advancing these ideas. People have to become more knowledgeable and engaged in the kind of complex issues associated with food production and sustainable agriculture in the US before they can participate in the kind of changes implied within sustainable agriculture.

Table 7.4 *Continuous and Lifelong Learning Through Schoolyard Organic Gardens*

Constraints	*Opportunities*
• There is no available land. • School year does not coincide with the growing season. • There is a high lead content in soil. • There is often no budget available. • Lack of support or understanding by decision-makers acts as a constraint. • Teachers often don't know how to do it. • Regulations regarding food use and consumption in schools. • Is there a lack of people to initiate, develop and supervize the project?. • There is a feeling of frustration if projects are not followed through.	• Tear up the asphalt. • Link with year-round school curricula. • Build a greenhouse and hold classes in it. • Take the school to garden/farms. • Children learn how to garden. • Learning creates empathy with conventional and organic farmers. • Offset food costs for school – create a healthy school lunch. • Group/community building – a garden will create and sustain teamwork. • Children will learn about time management and personal responsibility. • Teach intergenerational stewardship – hand over the soil in better shape to younger children. • Develop curriculum around the garden experience: science, geography, cooking, time management, history, math, reading/writing. • Help children learn that you don't just judge food by price and appearance – the quality of food is more than that.

Government action and regulation

Examining the role of government in supporting change at the pesticide and sustainability boundary was critical, but a number of contradictory lines of discussion arose. On one hand, it was felt that lobbying from industry led to a less responsible government response to the issue. But regulations were also seen as 'sticks' rather than 'carrots' which could be backed by funding for the development of alternatives. The process through which regulations are established was criticized. Regulations frequently represent election-year political pressure and a response to crisis decisions, rather than responses to public discussion of need and benefit. There are very few forums for such a process.

The role of the farming community in the development of regulations was raised. It was felt that there is not enough self-discipline among farmers to eliminate regulations. Farmers' antagonism to regulations is often based on their own agricultural practice. One statement put it this way: 'If I was a practising environmentally aware farmer, I would want government regulation to be in place and universally enforced.' A final paradox reinforces the sense that this area is rarely discussed in multi-interest forums. Two sides of regulation were identified: the 'dark-side' which stifles innovation in contrast with the sense among others that regulations have 'dragged' farmers into sustainability and made them more innovative.

Role of local communities

In common with many of those searching for answers to complex problems, the participants at this event viewed local community as an important locus for change. The point here is well known and, interestingly, backed by both ends of the political spectrum – albeit for different reasons. While some feel that support for increased community responsibility is merely rhetoric which rationalizes reduced spending by federal and state governments, others believe that stable rural communities, which are economically, socially and environmentally sound, keep more money in the community, encourage more food to be produced and sold locally and emphasize the capacity to sell services locally.

This approach obviously has close connections with such concepts as bioregionalism and sustainability. The idea would be to 'encourage cultural and ecological diversity across the landscape with the aim of stimulating innovation'. The bioregion was viewed as an appropriate means for defining 'community' (eg based on the sharing of a common watershed). This would further stimulate the development of many small-scale regional processes which might collectively gain access to the large organizations that dominate the agricultural scene. In assessing the constraints and opportunities of this kind of enterprise, frameworks were developed as a basis for establishing potential areas for action by one of the small working groups (see Tables 7.5 and 7.6).

The food security concept and its connection with healthy foods and local markets was also connected by participants with the community theme. The point was made that encouraging food security in the form of a stable and healthy food system involves incorporating all the costs in the economics of food supply (health, environmental and social). This implies upgrading and optimizing local market potentials, enhancing the visibility of local labels and developing consumer awareness of food origins and production methods. Participants believed that this local community approach to supporting indigenous crops and farming implies a recasting of the way we look at food issues through reviewing the entire chain of production process and distribution.

Table 7.5 *Strong local communities for rural America*

Constraints	*Opportunities*
• Need to add value to raw products at the local level – processing should be more encouraged at the local community level. Presently there are insufficient subsidies for these activities. • At present the financial system (state and national) doesn't support people entering farming – no incentives. • The present tax system does not allow money and property (transfer of assets) to be passed on to the next generation without penalty.	• Foster a healthy number of local farmers; other small local businesses to supply goods and services; and locally based institutions, churches, and schools to keep small rural communities viable (money remains in the community). • Cooperatives offer a possible vehicle to strengthen local growth. • Stronger local communities will lead to a better quality of life for all Americans – education, social life at a local level, better connections with the rest of the world, a unified effort to aim towards cleaner air and water and better soil stewardship.

Table 7.6 *Supporting local communities using a bioregional perspective*

Constraints	*Opportunities*
• The present education system does not adequately cover food production and associated issues or environmental sustainability at local, national, and global levels. • The current mind set – with respect to planning, people have not made the leap to thinking in terms of bioregions – we still tend to think in terms of existing small towns, suburbs and large cities. • The present political system is not organized in a way that encourages a bioregional perspective. • Economy of scale (capital investments required, mobility of capital mobility of raw products) encourages large-scale production for larger profits. • Public ignorance about the complex issues associated with sustainability – confusion and frustration experienced by people who are uninformed.	• Reinvent a framework for the future that applies traditional values in a modern context. • There is the potential to create interconnected systems organized within bioregions (for instance, agricultural, information, communication) that could support locally based sustainable development. • Public awareness – promote the idea of buying locally whenever possible in order to support local industry and community-supported agriculture (for example, farmers who are loaned money from local sources within the bioregion). • Create a political party that would support environmental sustainability and a bioregional perspective to development. • Use further search processes to involve people directly in food production – to engage them in the process begun with this event.

Creating new institutions and change processes

It was recognized throughout the future search event that virtually all of the change processes proposed to connect alternative agriculture with the pesticide industry require institutional contexts that do *not* now exist and methods that are not generally applied at present. This section presents an outline of the networks, dialogues, taskforces and other forums that future-search participants considered necessary in order to establish a common ground for generating action on *all* elements of the problem. This implies organizing for continuous change in the future by bringing together people with different interests but mutual concerns.

Because of the depth, history and passion that marks the pesticide and sustainability dichotomy, it would be naive to expect complete consensus regarding future directions out of a two-and-a-half-day engagement. The future search was chosen as a means of triggering change in the historic patterns of debate and confrontation. The organizers and participants understood that they were embarking on a long journey. Some very satisfying first steps were taken; seeds were sown to begin a broad-based process of inquiry and adaptation. It was pointed out that a danger of such an enterprise is that it will eventually return to the 'continued spinning of crippled discourses' in which there is a collision between the claims of chemical companies versus a shared vision of the future as a dialogue of trust.

Considerable discussion addressed this dilemma of creating trust. It was felt by some that taking positions on the basis of environmental impact demands a capacity to know and understand subtle differences in the effects of particular actions. In addition, the positions taken by corporate lobbyists on the one hand and environmental activists on the other often sharpen the boundaries between competing opinions and, therefore, reduce trust. It was suggested that we need new ways of building trust (for instance, the development of joint research projects involving both business and environmentalist interests). In such a context 'trust means introducing civility into the debate ... conflict is not negative if it leads to the crossing of boundaries as opposed to character assaults'. This civility issue seemed critical to one group of discussants – 'when debating issues in conflict ... we must make the assumption that we are all decent people and that one institution is as valid as any other'. But this also implies a measure of social balancing which is able to bridge the gap between trusting individuals who are being dealt with directly and trusting collectives, such as companies.

Participants felt that if action is to be taken at the local level, it is important to create intermediate institutions to address those issues which relate to the needs of particular places by creating relevant processes and events. It is also necessary to create settings to accommodate the array of players and interests brought together at that time. It was suggested that a common principle of this activity should be to consciously press for the engagement of relevant organizations in a *system-wide dialogue*. Some believed that it is essential to think of these sessions as forms of exploratory conversations – discourses aimed at searching for different ways of dealing with the inherent conflict by adopting, in the words of a participant, 'a reparative way'. This area formed a critical field of action for many of the participants who supported the further use of search-style processes as a means of fostering democratic engagement with the notion of bioregionalism in the context of food production and sustainable agriculture. Such a process, however, should begin with one-to-one conversations with leaders of communities and key figures who 'have the ear of the people' – individuals with credibility at local, state and national levels.

Two further potential areas for action were proposed by small working groups in the constraints and opportunities stage of the search process, focusing on this theme of facilitating environments of trust and encouraging local ecosystem-based activity (see Tables 7.7 and 7.8). Also, one of the final task groups applied many of these principles with regard to the development of locally based watershed management bodies. Initial discussion on creating watershed management authorities took place during the constraints and opportunities phase. One working group pointed out that the focus of such operations is ecosystem integrity expressed in terms of watershed conditions. The argument for change and the catalyst for action arises from the pollution of surface water resulting from pesticides and nutrients. This is an important unifying issue that has a clear local relevance. The participation of farmers has to overcome their overall rejection of anything associated with imposed standards and administrative threats. Watershed management authorities would have to be based on trust and demonstration of the proven success of such initiatives.

It is important to establish involvement from the beginning. Voluntary action and peer pressure are important elements that encourage involvement. Also, the objectives have to be seen to be associated with a potentially serious situation that demands attention and the incorporation of a future collective vision. The quality of local leadership is of major importance. The momentum of such activity builds from success. An important stimulus of support for this holistic approach to sustainability is the demonstration provided by 1000 farmer-led watershed associations, many of which already involve the participation of chemical companies.

Table 7.7 *Creating a facilitative environment of trust among agricultural companies, consumer groups, governments, farmers, and other stakeholders*

Constraints	*Opportunities*
• Trust is earned and not produced. • Earning trust takes time, effort, and self-confidence. • It is easy to appear to be deceptive while trying to build trust. • It involves a perceived loss of power on part of agricultural companies. • One bad misstep can cause the process to collapse – trust is fragile. • Trust leads to increased vulnerability. • Many companies fail to legitimize either personal or employees' ethics.	• Communicate with customers and involve them in new directions. • Process can generate new ideas and more creative ways of working. • Getting beyond 'agreeing to disagree' can generate breakthroughs. • Job exchanges build understanding and knowledge. • Need to help people to reconcile their ethics and the organization's ethics. *Methods:* • Future search and other processes are designed to build trust. • High-level consortium of top leaders lead their organizations into change processes.

Table 7.8 *Local-level ecosystem-based activity*

Constraints	*Opportunities*
• We do not appear to be making progress. • We leave the grassroots out. • It is difficult to initiate process where issues are addressed before a crisis develops. • There is a lack of monitoring of changing conditions to generate interest. • Local initiatives alone are not successful – they need financial and political support. • It is difficult to raise interest in apparently stable and secure areas where there may be a complete lack of concern regarding such issues as soil erosion. • It is difficult to involve chemical companies unless there are direct changes affecting them.	• Recognize the need for whole-system thinking – draw on a bubbling-up effect from all levels. • Emphasise unifying theme – such as water – connect fishing, farming, hunting, consumers (urban and rural). • Important to raise issues that demand attention – situations in which people are invited to be involved from the start of the process • Local leadership, voluntary action, and effect of peer pressure are important. • Momentum is built on success (social, political and environmental). • Have to build on interest relating to conservation/wildlife aspects of sustainable agriculture. • Actions initiated locally will often take tougher positions than government regulators – this results in better cooperation in monitoring and enforcing regulations.

Selection of Task Groups

While the primary objective of this particular future search process was exploratory, an opportunity was provided for participants to select specific areas in which they believe action should be taken. The way in which the set of potential action steps was developed provides an insight into the overall perspectives of the participants. An examination of the array of issues requiring action and the final selection of the task-group themes provides a basis for reviewing the outcomes of the event. The close paralleling of the task issues and the discussion themes is noteworthy (see Table 7.9).

Selected task groups were:

- Stewardship I;
- Stewardship II;
- Technology;
- Watershed Bodies.

A second task group focusing on stewardship was established because of the large number of participants who wished to use this concept to conclude their activity in the future search. The fact that half of the participants wanted to work on this topic suggests that an integrated response is an essential element of progress in the area.

Stewardship 1

This group defined stewardship as a key integrating notion which implied 'leaving land and resources better than we found it; a long-term perspective on preservation, conservation and regeneration'. The primary interest was in encouraging and supporting the absorption of these principles into business settings. Stewardship was regarded as a philosophical and moral issue which can bridge local and global aspects of sustainability concerns – 'think globally, act locally'. The following objectives frame the task group's notion of stewardship:

- a balance of business, culture, and ecology;
- an innovation demanding new infrastructure and alliances;

Table 7.9 *Participant definition of potential task group themes*

- search process that focuses on democracy and bioregionalism in context of sustainable agriculture – role of political advocacy;
- alternative farming experiments;
- improve education about food production/sustainable agriculture in the public school system – continuous experiential learning about food production for children;
- public awareness/education about bioregionalism;
- public awareness and participation generating greater involvement of the public in sustainable agriculture;
- stewardship – balancing values of business, culture and ecology in environmental, economic, regulatory and social context;
- regulations based on need, current information, and research;
- benefits/impacts of technology;
- foster locally based, holistic, system-wide efforts to manage watersheds.

- a new management philosophy based on sustainable natural systems;
- a transferable process which business can use to gain returns on investment.

The task group chose to follow-up its objectives through developing a prototype organization which would introduce stewardship to management. The following specific tasks were defined:

- Establish a prototype organization (the task group called it Ecosystemic Change Incorporated) designed to initiate and introduce stewardship approach into business organizations.
- Create a team of experts to inform business about opportunities on multilevels (organic farmers, entrepreneurs, processors, distributors, industry, experts, retail, etc).
- Use initial examples as 'incubators', exploring profit and non-profit opportunities and introducing the idea to existing organizations.
- Form management teams able to create new infrastructures that reflect an overall management approach to stewardship in organizations.
- Encourage the addition of projects that are based on a real-world appreciation of business stewardship to business school curricula.

This task group treated the idea of stewardship as integrating many of the change perspectives developed during the earlier stages of the future search. In this respect the group took concepts that appeared to be counter to established business theory and practice and proposed a means of translating the future search approach into an operating organization.

Stewardship 2

The second stewardship task group covered similar ground to the first by considering ways in which to increase the numbers of businesses that promote sustainable practice. For the members of this group, corporate stewardship implied the developing of long-term ecologically responsible ways of doing business.

Several strategies were considered to achieve the objective of fostering corporate stewardship:

- Convince shareholders of public companies that there are long-term gains to be made by investing in firms committed to sound ecological practices. Provide ways of demonstrating the achievements of successful, profit-making 'green businesses' as an important aspect of supporting ecologically sound business practices.
- Demonstrate that full-cost accounting is an important means of demonstrating the issue of stewardship to businesses, legislators and the public. This demands the use of intensive lobbying of legislatures to accept new products according to their real-costs (costs related to health and the ecosystem).
- Engage the general public in encouraging stewardship through making people aware of the business interests underlying investment and pension funds (making contradictions in current investment practices more visible – for instance, pension plans of medical associations investing in the tobacco industry). Make people more aware of where their money is invested and promote 'green' mutual funds.

- Provide more knowledge to consumers regarding 'greenwashing practice' by encouraging them to ask what companies make and how their products are produced. Promote the idea of 'sustainable consumption' (a movement in Europe which targets new business and consumer ethics in processes aimed at sustainable practice).
- Develop the means of encouraging stewardship as a primary management principle for the future through the use of search-style events to involve participants from a specific bioregion in initiating sustainable practices, especially by targeting practitioners (those directly engaged in food production).
- Use search-style processes for engaging young people in issues related to ecologically sound business, sustainable consumption and progressive investment practices. Experiential learning initiatives, such as gardening in schools, could be drawn into search processes as a means of encouraging involvement in sustainable agricultural practice.

Appropriate Technology

This group chose to address the virtually 'unaddressable' issue of the search for appropriate technologies that are in tune with both the pesticide industry and sustainable agriculture interests. The challenge of dealing with this issue haunted the proceedings of the future search. The discussion leading to the formulation of this task group topic demonstrated the problem by highlighting the orientation of some participants on the 'limits and impact' of technology as opposed to others who felt the focus should be the 'benefits and opportunities' of technology. Agreeing on the objectives of this task group was, in itself, an exercise in creating an environment of trust, as discussed earlier.

The direction of the discussion was characterized by the point that: 'this is not an issue of chemistry but of information – of labels and regulations'. Nevertheless, the group agreed that sustainability necessitates searching for chemicals that have narrow targets and minimum impact rather than a search for 'stability without insecticide'. Considerable time was devoted to understanding the nature of the balance between technological choice and social outcomes. It was felt that there is a need for wider participation in technological assessment and choice with the aim of creating a wider understanding of the trade-offs and who gets them – a democratization of technology assessment. Here the assertion was that 'the problem is not toxicological, but process'. The basic need is for 'a democratically based ex-ante technological assessment process which is widely used'. An important question was: 'Does this all still come down to the difference between biological and chemical systems of control?' The task group carried out a brief exploration of this distinction with reference to the case of a hypothetical soil pest attack on a potato crop.

Making distinctions between biological and chemical systems generated yet more questions for the task group:

- Is there a middle ground that minimizes risks – a third way?
- Does this decision have anything to do with the consumer? (They currently only express themselves through the market and there is a general lack of information available on research.)
- Should research results be open to the impact of informed public choice?
- Who is advocating on behalf of the consumer?
- Does the chemical industry have a role in the wider information system (versus merely delivering technology)?

The group concluded by considering the outcomes of the discussion. They felt that they had gained a clearer understanding of where the industry and toxicologists are coming from, and that both industry representatives and consumers benefit from such a discussion. But there remained a strong feeling that, more than anything else, 'the discussion emphasized the significance of what we don't know in the making of these decisions!'

Watershed Management

This group continued the discussion of watershed management initiatives and focused on how to foster locally initiated, holistic, regional, water-oriented associations. The key issues raised reinforce much of the previous discussion:

- selecting the type of process required to generate change;
- engaging and integrating all interests in the process;
- selecting leaders (preferably not governmental);
- ensuring that potential solutions are economically viable and feasible;
- incorporating trust to unite the process;
- making the process flexible, open, and self-regulating.

Conclusions

Much of the discussion at this first future search conference on pesticides and sustainability was concerned with the framing of issues among the varied group of participants, but there was also a pervading sense of the need to create strategies to establish alternative futures for this vital area. In framing the problems associated with this area, the participants always recognized the fundamental divisions between people representing competing paradigms for the future of food production in the United States. However, the debate constantly returned to the middle ground of perspectives on which a whole-system design for the future must be based. It was necessary to establish processes that engage the wide array of actors who should determine future directions – 'we are all in the same crisis situation'. The conclusion was that no one can deny the seriousness of the problems facing us on a global to a local scale and that sustainability, combining both food security and environmental protection, incorporates the directions in which change must be taken.

There are many proposals for action coming out of this future search event, but few of them are directly concerned with pesticide use and the pesticide industry itself. This is mainly the result of decisions made by the participants to explore the wider context of the sustainability and pesticide issue. They chose to focus on the forces limiting the extent of change and the ways in which such forces could redeploy themselves. However hard it is to generate fundamental change among established institutions, this future search discovered a number of paths which, it was generally agreed, urgently need to be followed.

Central to this discussion was the dilemma of maintaining supplies of relatively inexpensive food (which the current agricultural system associates with the application of chemical pesticides) *while* at the same time responding to the fears of toxicity and its impacts on human and non-human environments, recognizing that these are as much social and political considerations as they are technical and economic. The dominance of the established pesticide industry position and the acceptance that chemicals are still the cheapest and simplest solution to the production of food suggested to this group that major change is not going to take place solely *within* the agribusiness system. The problems being addressed have basic implications for the underlying quality of life for all Americans over the long term. This prompted the future-search participants to spend a great deal of time and energy focusing this debate on the broadest possible societal arena. Public awareness, education, community action, and creating new forums for discussion and public action were all central aspects of the strategies proposed.

There is no doubt that the consensus emerging from this event was that sustainable agriculture is an important symbol of the process of facilitating change. It was seen not only to demonstrate the full implications of applying an alternative 'stewardship' to agricultural practice, but also to draw attention to the principles of full-cost accounting, environmental sustainability and bioregionalism, all of which must become central elements of change for industry, government, and consumers alike. But the primary concern recognized by the participants was clear. How can one initiate and maintain processes of change that engage all critical elements of the domain? How can one create a middle ground arena that takes the discussion from extreme positions and confrontational stances regarding sustainable agricultural interests and the pesticide industry? The future search, itself, was a model of the kind of alternative setting which allows the divide to be at least temporarily crossed.

The proposals for future action that were made throughout the search event deal with several levels of interaction. The *community* is seen as an important change setting, particularly where an increased control over local food systems can be demonstrated and where the ideas of food security, healthy consumption, and producer and consumer partnerships can be played out in real time. There is potential for the *bioregionalism* concept to become a bridge between local rural and wider metropolitan expressions of these ideas. The importance of generating *public awareness* was constantly raised at the future search. Again, this was viewed as a means of connecting a series of public concerns: environmental health, shareholder activism and the investment in 'green' mutual funds are all active public interests. The participants believed that what is needed are forums and opportunities for action that help overcome the pervading sense of consumer helplessness in the face of threatening, but apparently unchanging, institutional systems.

Education at all levels was seen as an essential element. It must involve all parts of the domain of interests: schoolchildren, consumers, growers, legislators, industrialists. Children were seen as being integral to this process and several sessions called for 'organic gardens in every schoolyard'. But knowledge is seen as being tightly controlled at present and considerable discussion emphasized the need to make information on toxicology and technology more accessible to the public – 'a democratization of technology assessment'. This also links with the importance of providing open access to research, which was perceived as being controlled by the pesticide industry. However, in relation to this entire issue, the point was constantly made that 'the problem is not toxicological, rather it is one of process'. Therefore, the central ongoing proposal coming out of these sessions is that maximum effort is needed to create new forums for discussion which cross the complex, differentiating boundaries discussed here. This means that it is critical that we establish networks, dialogues, taskforces and other formal and informal organizations for continuous change 'which bring together people with different interests, but with mutual concerns'. The importance of such strategies to this group of participants is demonstrated by the list of task topics that were generated in order to select the themes of the concluding task group activity (see Table 7.9). The issue of trust is central to such interactive processes; nevertheless, this is a quality that cannot be generated out of rhetoric but only from successful experience. Creating settings in which people from all positions can discover such experience seemed critical to this group. Clearly, this is exactly what they gained from this future search event.

REFERENCES

Emery F (1993) 'The Second Design Principle: Participation and the Democratization of Work' in Trist E and Murray H (eds) *The Social Engagement of Social Science*, pp 214–33. University of Pennsylvania Press, Philadelphia.

Emery F and Trist E (1973) *Towards a Social Ecology: Contextual Appreciations of the Future in the Present.* Plenum, London.

Franklin B and Morley D (1992) 'Contextural Searching: Case from Waste Management, Nature Tourism, and Personal Support' in Weisbord M (ed) *Discovering Common Ground*, pp 229–46. Berrett-Koehler, San Francisco.

Morley D and Trist E (1992) 'Planning, Designing and Managing Large-Scale Searches' in Weisbord M (ed) *Discovering Common Ground*, pp 187–214. Berrett Koehler, San Francisco.

Morley D and Trist E (1993) 'A Brief Introduction to the Emerys' Search Conference' in Trist E and Murray H (eds) *The Social Engagement of Social Science*, pp 674–78. University of Pennsylvania Press, Philadelphia.

Weisbord M (1992) *Discovering Common Ground: How Future Search Conferences Bring People Together to Achieve Breakthrough Innovation, Empowerment, Shared Vision, and Collaborative Action.* Berrett-Koehler, San Francisco.

Weisbord M and Janoff S (1995) *Future Search: An Action Guide to Finding Common Ground in Organizations & Communities.* Berrett-Koehler, San Francisco.

Wright S and Morley D (eds) (1989) *Learning Works: Searching for Organizational Futures.* ABL Publications, Faculty of Environmental Studies, York University, Toronto.

8

SOLVING FOR PATTERN

*William Vorley and Dennis Keeney**

The less you know about sustainability, the better it sounds.

Robert Solow

LEVERAGE POINTS

The past seven chapters have examined the pesticide business in the context of sustainable agriculture and the greening of industry. The journey has been discomforting. Our preconceptions of business and science feeding the world, or of uncaring business empires despoiling the earth, have been rattled and shaken. We are obliged to redefine the basis on which we recommend changes to the system, be they changes by business, by policy-makers and government agencies, or by individuals. Clearly, we are looking for leverage within *composite problems* (see Clayton and Radcliffe, 1996) at all levels – industry, agriculture and ecosystem. Because of the complex systems that underlie composite problems, there are repeated instances of unintended and sometimes undesirable consequences when companies, farmers or government agencies take a direct and linear approach to solving those problems. This gives us advance warning that the most powerful leverage points could be counter intuitive to industry and its regulators. If we are to address the *causes* rather than the symptoms of these problems, we will have to think beyond efficiency, towards *redesign* (MacRae et al, 1993).

Our task now is to highlight those leverage points, building recommendations from industry's own language. In doing so, we are attempting to move the debate from *issues management* to a *strategy for sustainability*. In other words, we wish to move from a problem-centred look at probable futures (how will sustainable agriculture affect my business?

* Apologies to Wendell Berry, whose essay 'Solving for Pattern' (in Berry, 1981) is an inspiration.

how can we limit the pollution from agriculture?) towards a mission-directed look at preferred futures (how can my business or my agency contribute towards achieving agricultural sustainability?) (see Ellyard, 1994). We first summarize the forces which are shaping the pesticide industry and its markets, and then look again at industry visions for sustainable development. We take stock of the implications of those visions, especially in terms of the ecosystem concept. With those impacts in mind, we revisit the leverage points identified at the beginning of the book to explore their potential for turning ideas into action, for walking the sustainability talk.

THE STORM CLOUDS

The serious threats to the pesticide industry's profitability are described by Koechlin and Wittke (Chapter 5) and by Pretty et al (Chapter 2). Some of these threats are a direct result of 40 years of success – a maturation of the pesticide market, with pressure on prices from generic producers and a realization that many established pesticides are now commodity rather than speciality products. Then there are the economic trends in agriculture which were indirectly triggered by a host of technologies, including pesticides, that ushered in the 'efficiency' revolution in agriculture. These technologies have for decades been moving people, profits and political bargaining power out of farming to other parts of the agrifood chain. This shift in power has strengthened the role of food processors and retailers in farmers' decision-making. Then there are the wider trends of globalization of trade and deregulation of agriculture (part of a larger corporate agenda for expanded opportunities in trade and commerce), which are dismantling production subsidies and encouraging global sourcing of food from the lowest-cost producers. And, finally, there are threats from biotechnology, which will likely shift large and lucrative chunks of the pesticide market from chemical- to seed-based products, which manufacturers are anticipating through expensive acquisitions of seed and genetics companies.

Business leaders have added a further complication to this stormy business climate – a vision for sustainable development. Is this an opportunity or threat? Monsanto's Shapiro thinks opportunity. He sees the premise of the agrochemical industry becoming obsolete as the business moves through the critical discontinuity of sustainable development. 'Business grounded in the old model', says Shapiro, 'will become obsolete and die' (Magretta, 1997). But Shapiro's perception of this critical discontinuity or *breakpoint* is part of a future in which biotechnology rescues society from the impossible task of producing more food from less land. His gamble hinges on steering his company through this breakpoint by catching the 'second curve' of biotechnology: close down chemistry research and milk the current portfolio of chemical pesticides as 'cash cows'. Invest heavily in genetics and market access for seed-based technology. Focus on biological markets, while divesting basic chemical and industrial divisions. This is certainly the direction taken by the big players. Monsanto has invested nearly $2 billion over the past two years on access to biotechnology and seed markets (Lenzer and Upbin, 1997), and Novartis Seeds is spending in excess of $100 million a year on biotechnology.

Suppose, however, that the theory of pesticide companies is becoming obsolete for other reasons, and that there are other breakpoints in the system which bode less well for the sustainability of this research-intensive industry? Consider the following alternative scenarios:

- Agricultural producers, realizing that their dependence on chemical pesticides has only trapped them on a technology treadmill, and that public hostility is weakening their political position, and that biotechnology will only widen the rift between farming and public, decide that farmers stand to benefit more by moving their political support from the input sector to the consumer sector. That would leave the pesticide industry exposed to strong legislative pressure, and would make the company line about 'we in agriculture' ring hollow indeed.

- Biotechnology starts to look like an environmental liability in the eyes of shareholders, with companies being sued for the creation of aggressive new strains of viruses or weeds. That would leave the industry, which already has significant liabilities from the chemical era in terms of contaminated production sites and issues of worker health, looking less like an ideal investment.
- Shareholder activism or new legislation really starts to affect companies' chemical cash cows and the ability to fund R&D for new biotechnology products.
- New social movements rapidly bring latent consumer concerns out of the closet, in a decisive market shift towards organic or near-organic produce (see Hamilton, 1996).

Our point is that a vision for sustainable development is only as good as the insight of its creators, and its implementation is only as good as the insight and 'literacy' of a company or agency's employees. As the saying goes, if your only tool is a hammer, you will view all your problems as nails. A company or agency with a technocentric worldview will use a technocentric vision of sustainable development, in which environmental threats are seen as management problems with managerial and technological solutions. An oversimplified reality and uniformity of perspective may leave an organization vulnerable to unforeseen breakpoints. Organizations must first seek *literacy* before rushing to apply more 'sustainable' technological and managerial fixes.

Change Course or Change Colour? New Literacy as a Prerequisite

Without the perspectives which literacy can unveil, there is no sound basis for innovation. Literacy is the prerequisite for a company or an institution to fully understand its *stake* in the system, and thereby to decide which of its interests are vested in that system. It is the prerequisite of understanding the real threats and opportunities to your business. If a company learns that it can redefine its role from a chemical company to a pest management company, an agricultural company, or even an 'ecosystem management' company, it could drastically reduce the company's need to defend chemistry. Internal learning has a relatively low cost compared to the expensive mistakes of tunnel vision, which have seen companies scrambling to relearn and to repair their image.

How can corporate literacy be achieved outside of the rarefied air of the boardroom? Companies can follow the lead of Monsanto, inviting the gurus of sustainable agriculture and new business, such as Wes Jackson and Paul Hawken, to their headquarters to set management teams on the right path of applying the company's technological skills to meet perceived global needs. Monsanto has seven sustainability teams totalling 140 staff members: the eco-efficiency team, the full-cost accounting team, the index team, the new business/new products team, the water team and the global hunger team (see Magretta, 1997). Or companies could take the DuPont route which has a slower pace of experiential learning, including a well-designed series of field trials which compare conventional agricultural systems with the best that reduced-input and organic agriculture can offer. These trials are an open collaboration with specialists in organic farming, such as the Rodale Institute, and seem to have successfully avoided the antiorganic sentiments of agribusiness. The French-based company Rhône Poulenc has also experimented with organic farming.

Or, of course, companies can stick to the credo of *Saving the World with Pesticides and Plastic*, continue translating sustainable agriculture into defensive endorsements of the status quo, keep up a stream of propaganda which puts pesticides at the centre of efforts to 'feed the world', and maintain their all-out defence of chemical products until 'sound science' shows significant proof of harm. They can continue to press for 'enlightening the consumer' rather than enlightening their own personnel. But this position consumes much of the political and institutional capital of both industry and environmental groups (Wolcott, 1993).

Our proposal is for a literacy campaign which offers a more profound and personal challenge to the institutional mind-set, in order to explore the potential of *redesign* strategies. Innovations for efficiency or substitution come from within a company's own laboratories and field stations, or from universities and public research. But for innovations in redesign we have to look to the pioneers and iconoclasts – farmers, researchers and advisors who are ahead of the curve in shifting from a consuming to a conserving agriculture; who are getting high yields and regenerating their natural resource base with few external inputs; who are refusing to convert source into resource. We propose that pesticide companies *and* regulatory agencies such as the US-EPA develop a goal for at least 5 per cent of middle management to spend sabbaticals of up to one year on the farms and research stations of innovators in alternative agriculture.

There is at least one lesson from history which emphasizes the potential benefits of literacy, for companies to stay ahead of issues rather than being forced into belated, reactive and defensive postures. The emergence in the late 1960s and early 1970s of a more ecosystemic and rational approach to pest management under the umbrella term of integrated pest management (IPM) was treated by industry as a nonsense or as a threat to current business (van den Bosch, 1978, pp 173–4). This sowed the seeds of a view – still widespread today – that the agrochemical industry is obstructive to IPM implementation, or tends to distort the term towards integrated pest*icide* management (Thrupp, 1996, pp 24–9; Dinham, 1993, p19). A literate company with an ecosystem approach would have realized much earlier that IPM was in the long term interests of everybody, even if the overall market size for chemicals might shrink to a more ecologically and economically sustainable level. IPM opens market opportunities for high-value products that selectively affect pests and not their natural enemies. The need for selective products was recognized as early as the 1940s when biologists such as Wigglesworth (1945) and Ripper (1944) started to warn of the dangers of pest-control regimes built upon chemicals which were destructive not only to the pest but to the naturally occurring biological control agents such as predators and parasites. Even back in the early 1900s, cotton entomologists in the United States had warned of the dangers of unilateral approaches to pest control. And yet IPM was not seen as an opportunity by industry until the 1980s, and then only for the development of new highly selective products for niche markets, especially fruit crops, where the pesticide treadmill was leaving growers with fewer and fewer options. There were hardly any moves to review the existing product range, product usage recommendations or marketing strategies in the light of the IPM approach, though Novartis' Farmer Support Team has made real progress in this respect (*Farm Chemicals International*, November 1995).

The most salutary lesson of the IPM experience, and the core of a redesign approach to sustainability, is the importance of the *ecosystem concept* for understanding both the business of agriculture and the business of business. Literacy in ecosystems is the foundation for sustainability, because it is the means by which we see ourselves as contained within ecosystems, rather than as businesses or individuals that interact with environments (see EPA et al, 1995).

CORPORATE VISIONS AND THE ECOSYSTEM CONCEPT: IMPLICATIONS

An ecosystem is a complex of organisms and environment forming a functional whole; each species is dependent on the rest of the ecosystem for its maintenance. The concept of *interdependence* enshrined in the ecosystem approach is already part of corporate visions for sustainable development. Ciba-Geigy's Vision 2000, launched in 1989, aims to strike a balance between three 'equal-ranking' responsibilities – the economic, social and environmental – and thereby to 'ensure the prosperity of our enterprise beyond the year 2000'. This vision is remarkably similar to the ecosystem approach adopted by the Great Lakes Water Quality Agreement which operates on the principle of interdependence between society, economy and environment, and states that 'no segment of the circle can be sacrificed and all are essential to maintain a functional and sustainable ecosystem' (Hartig and Vallentyne, 1989; EPA at al, 1995).

It is fascinating how visionaries view the survival of a corporation and the survival of the Great Lakes in similar systemic terms. The corporation acknowledges its dependence on society (and acceptance by that society) and the environment for its survival, and acknowledges that human, natural and physical capital can no longer be thought of as easily substitutable. There are elements of both the *holism* which has characterized most systems thinkers in agriculture such as Aldo Leopold and Baird Callicott, and the more reductionist *systems dynamics* developed by engineers such as Jay Forrester and applied to social organizations by Forrester, Peter Senge and others (see Thompson, 1995, pp 118–46). The power of (eco)systems thinking is not just an appreciation of interdependence. It also puts us into the 'endogenous viewpoint' – the vantage point of viewing the *structure* of a system as the *cause of the problem behaviours* that it is experiencing (Richmond, 1994). It allows us to choose an entry point from which we can gain an understanding of why complex systems (large companies, lakes or farm fields) respond to management in the way they do, and an appreciation of the potential for new approaches to leverage agribusiness, legislators and farming towards more sustainable behaviours.

What are the implications of a systems approach for the management of industry and agriculture? We propose that a systems approach changes the way we apportion blame, the way we interpret effects, the amount of precaution and 'ecological room' which we include in our actions, and the relative importance which we give to environmental management and ecological design.

We're all insiders

There is a tendency for the advocates of systems thinking to sanctimoniously tut-tut the hopeless actions of the linear-thinking reductionists, locked in their feedback loops and blind to the obvious escape routes from their treadmills. This emerged clearly from the study by Mayhew and Alessi (Chapter 6):

> *Sustenance frequently expressed, with great exasperation, the lack of systemic thinking evidenced by short-sighted decision-makers. In doing so, they became selective of who and what is considered to be part of the system. They neglected the fact that there are people in the dialogue with competing needs and desires (for example, manufacturing, banking, government). They can't simply be dismissed from the system once identified as the enemy of their cause.*

So blaming outsiders – the press, environmentalists, the multinationals – is the antithesis of systems thinking. The comment by the Great Lakes project that the ecosystems approach 'relates people to ecosystems that contain them, rather than to environments with which they interact' means that humans are no longer seen as standing apart from the rest of the system, especially the agricultural ecosystem. Anthropocentrism, and definitions of sustainability grounded within the concept, are very difficult to marry with systems thinking. Being inside the system means that 'its desirable states cannot be defined *a priori* and solutions cannot be defined using methods (for example, those of the natural sciences) that rely only on an objective external viewpoint' (Waltner-Toews, 1996). We, along with the cause of our problems, are part of a single system (Senge, 1990, 67–78), which in turn tempers the tone of arrogance which environmentalists have used when talking to business. We see that what we do with our savings and pension funds, how we shop, how we eat and how we vote, all contribute to the pesticide issue.

Intensity, revenge effects and lateral solutions

Ecologists, agriculturists and business analysts are familiar with the symptoms of 'compensating feedback' when well-intentioned interventions call forth responses from the system that offset the benefits of the intervention (Watt, 1970; Senge, 1990). These interventions can leave the system fundamentally weaker than before and in need of further help. The harder you push, the harder the system pushes back. The war on pests has left us with a more virulent and more aggressive 'enemy'. For example, attempts to control forest pests such as the spruce budworm by spraying insecticide turned a sporadic problem into a semipermanent outbreak.

In his analysis of technology, Edward Tenner (1996) concludes that such 'revenge effects', or unintended consequences of technology, are the flip side of *intensity* – our propensity for head-on confrontation and our tendency to overextend a good thing, whether we are dealing with the automobile, with antibiotics or with pests. The cause of revenge effects is not a problem of technology as such, argues Tenner; technologies can also diversify and *de*-intensify. He calls instead for *finesse* in our technology, 'abandoning frontal attacks for solutions that rely on the same kind of latent properties that led to revenge effects in the first place'. This perspective of frontal versus lateral interventions offers us a useful way to review the more sustainable technologies that the pesticide industry is developing or already has on the market, such as genetically engineered crops with resistance to herbicides or insect pests. For example, could it be that Monsanto has constructed the problem in potato production to be the potato beetle, rather than potato monoculture or poor soil quality (Kloppenburg et al, 1996)? Is not the inevitable revenge effect of this intensive, nondurable technology a beetle resistant to bacteria as well as the array of chemicals to which it is already immune (see Robinson, 1996, pp 75–81; van Emden, 1991)? As another example, could it be that pesticide companies have constructed the problem in grain production to be weeds rather than the deterioration of the soil, which has required the compensating effect of massive increases in herbicide use between 1950 and 1980 (Ghersa et al, 1994; Benbrook, 1996, pp 18–20)? Could it be that weeds are actually an indicator of what's wrong with the soil (Hill and Ramsey, 1977)?

The fork in the road between linear and lateral integrative approaches to pest control was marked out long before systems-speak and computer modelling, and even before the widespread use of DDT in agriculture. In 1940, Sir Albert Howard published *An Agricultural Testament*, based on 40 years of field experience and observations in

England, the West Indies and especially India. He wondered why peasant crops were pest-free, and between 1905 and 1910 he set out to acquire the traditional knowledge, with the local farmers of Pusa, India, and the 'insects and fungi' as his instructors. After 30 more years of refining his ideas, Howard concluded:

> *The policy of protecting crops from pests by means of sprays, powders and so forth is unscientific and unsound as, even when successful, such procedure merely preserves the unfit and obscures the real problem – how to grow healthy crops.*

In an angry polemic against the pesticide-driven status quo, the noted entomologist and biocontrol specialist Robert van den Bosch (1978, pp 175–6) wrote:

> *[Researchers] consider their charge to be bug killing and simply do not understand that pest control is essentially an ecological matter: insect population management. Instead, with tunnel vision, they continue to seek simple answers... Historically, this simplistic approach has been repeatedly manifested in frenetic efforts to exploit innovations... Each innovation is seized upon as a potential panacea, heavily promoted, supported and researched, but none ever solves the insect problem. Indeed, they often prove counterproductive by diverting brains, energy and funds from integrative research.*

The lateral perspective can also provoke some interesting debate on some of industry's most ingrained philosophies. Take, for example, the premise that technology for increased production is the solution to world hunger; this is similar to the logic which led companies into commercially disastrous ventures in the 1970s with microbial proteins such as BP's 'Torpina' and ICI's 'Pruteen'. Could it be that Monsanto has constructed the problem of hunger as insufficient global food production rather than poverty, and that this world view has spilled over into their global hunger sustainability team? And could it be that industry has framed the problem of public disquiet with pesticides as a lack of proper unbiased information, rather than a deep and latent objection to chemical or genetic interference with the food supply?

Precaution and a new relationship with science

An ecosystems approach is *anticipatory* rather than reactive. It stresses the uncertainty of our knowledge and the possible 'irreversibilities' of our actions. That means maintaining flexibility and maximizing future choices by exercising restraint and leaving ecological room for manoeuvre. These are the precepts of the *precautionary principle* which is rapidly gaining acceptance in the arena of international environmental policy. A common misreading of the principle finds it Luddite and an impediment to technological development. Rather, the intention is to avoid the experiences of recent history, in which the time lag between evidence of harm and irrefutable proof of harm has been, in retrospect, ruinous. The chemical industry has preferred to 'play the trump card of uncertainty' rather than implement thoughtful action in advance of indisputable scientific proof. Demanding proof of harm betrays a misunderstanding of science, and cost-benefit analysis invariably favours industry. The costs of curbing a hazard are immediate and quantifiable while the benefits of avoiding the hazard are unprovable and hard to assess, and scattered among the population (Lean, 1996). The negative consequences for human and environmental health of the chemical industry's lack of precaution are well known, from lead in gasoline, through chlorofluorocarbons (CFCs), polychlorinated biphenyls (PCBs) and organochlorine pesticides.

How should pesticide companies now respond to the latest smoking gun of endocrine disruption, for which conclusive evidence may never be found, and subsequent issues of precaution versus proof: by piecemeal actions, or by a reappraisal of the principles of pest control? Pesticide technology is far from an ideal suitor for the precautionary principle, even when chemists and 'precision' application technologists have cleaned up their act. Pesticides are typical of chronic low-level environmental threats that fall through the epidemiological net. The very process of releasing potent chemical biocides into the environment to form unknowable and potentially synergistic combinations in the soil, water and air is inherently lacking in precaution. In a culture of precaution, pesticides and other ubiquitous man-made products are suspects for causing a whole range of unexplained maladies of the late 20th century, from breast cancer to deformed frogs, despite the continued use of science to claim innocence until uncategorical proof of guilt. This editorial in *Chemistry and Industry* (6 May 1996) is a wake-up call for industry:

> *If the industry believes its own tired rhetoric on sustainable development, self-regulation and better public relations, it will have to make more effort than it did with CFCs to respond [to the endocrine disruption issue] in a credible fashion. For one thing, it is critical to avoid even inadvertently suggesting that 'no evidence of harm' somehow equals 'evidence of no harm'. Equating the two, if only by implication, is just as misleading as the environmentalists' exaggerations.*

The precautionary principle changes the balance of power between science and the community, and its implementation is part of the new relationship between industry and science called for by Mayhew and Alessi in Chapter 6.

Ecological design and the restorative economy

There is a growing school of thought that ecosystems are not only the model for us to rediscover our interdependence, but also a blueprint for the design of our business operations. Companies have traditionally measured their environmental performance using management systems, and many pesticide manufacturers have reported impressive reductions in emissions from their production facilities. But commentators such as Paul Hawken (1993, p54) argue that if we are sincere about moving from an industrial economy (which Hawken likens to an immature ecosystem) to a restorative society, 'companies must re-envision and re-imagine themselves as cyclical corporations'.

An increasing number of commentators are voicing an opinion that we are perfecting means and confusing ends – that we have a *design* problem as well as a management problem (Hawken, 1993, pxiii). Richard Welford (1994, pp 185–6) believes that environmental management systems can 'take a business round in circle after circle of incremental improvement but, where the starting point for this environmental merry-go-round is fundamentally unsustainable, it is unlikely to result in the significant step-up in environmental performance which the world needs.' Environmental efficiency is like economic efficiency – a means to an end and not an end in itself. Welford proposes that firms audit for sustainability, which 'challenges firms to prioritize their actions in ecological terms rather than management systems terms. Moreover, it must be recognised that while bad systems may worsen environmental degradation, it is not the systems themselves which cause the damage but the products and services for which the system is designed.'

A chemical company that supplies products to agriculture may see some irony in the fact that one of the closest models to the ideal of a restorative or a replenishing economy is farming, or at least non-industrial farming. Wendell Berry (1978) and many before him have pointed out that in a healthy economic life, agriculture and industry form two mutually supporting poles: farming as prime production (creating resources and few wastes) and industry as production via processing (creating waste as a by-product). The arguments that biologically intensive cyclical agriculture needs to be subsidized with chemistry and biotechnology have been discussed in Chapter 2. But it was another push for closed-system process design – the 'Natural Step' – that stimulated the Federation of Swedish Farmers to take on the mission of 'moving toward the world's cleanest agriculture', aiming at a significant reduction in the use of chemical fertilizers and pesticides. Ecosystem thinking in agriculture focuses on plant health rather than pest death, designing and managing the agricultural ecosystem to intensify the cycling of nutrients, the natural control of pests and the conservation of soil (see Schönbeck, 1989; Altieri et al, 1996). It appreciates that environmental management systems for agriculture such as best management practices are, like their corporate equivalents, incapable of turning an inherently unsustainable farming system into a sustainable one.

An example of this is herbicide runoff and soil loss in cash-grain farming in the US midwest. Pesticide and soil loss can be reduced though changes in tillage practices and other best management practices, such as reduced herbicide rates and riparian filter strips. But if we look at the cash-grain livestock system from a perspective of ecological design, we could propose that instead of feeding animals with grains in huge feedlots, we could intensify livestock production on *grass*, a perennial crop famous for its soil, building qualities and low pesticide requirements.

To summarize, business visions for sustainability acknowledge the interdependence of business, society and environment. In this simple statement are profound implications for the way in which a business sees itself and its products, and the way in which it views new opportunities in sustainable development. Let us now look at the leverage points which were identified in Chapter 1 and developed in Chapters 3 to 7, from an ecosystemic perspective.

OPPORTUNITIES FOR SUSTAINABLE BUSINESS IN SUSTAINABLE AGRICULTURE?

We have seen in Chapter 2 that high-yield farming with fewer external inputs or with organic farming methods is possible, but requires a sophisticated blend of local knowledge and technology to be successful. By successful, we mean agriculture which restores the complexity and resiliency of the agricultural ecosystem, and which is biologically intensive rather than simply less chemically dependent. So, at first, there appear to be big opportunities for the private sector to make money and be consistent with their visionary talk in developing products and services for sustainable agriculture.

'Soft' chemistry: biological and biotechnology products

The obvious business opportunity for pesticide companies in sustainable agriculture is *substitution technology* – softer versions of current pesticide products. Companies have invested heavily, with considerable success, in developing IPM-compatible chemical, biological and biotechnological products for insect, weed and disease control which are active at low rates or which are selectively toxic to the target organisms.

If this book had been written just two years ago, we might have concluded that, despite the problems of small niches, short shelf lives and quality control, the pesticide industry was adjusting its research and marketing to accommodate biological products.

But as Koechlin and Wittke comment in Chapter 5, this 'biologicals' phase has been almost completely eclipsed by biotechnology. Companies realize that they can transition from broad spectrum, durable chemicals with expiring patents for the large agricultural markets to broad spectrum, durable and patented biotech–seed technologies for the same markets, with little change needed in corporate organization or philosophy.

An ecosystemic view can call into question the long-term market potential of these specific chemical and biological products. For insect control, it is ironic that most of the pest-control submarkets which are big enough for the industry to warrant the development of soft products, are *secondary pests* caused by a history of broad-spectrum pesticide misuse or other mismanagement of the agricultural system. Once those systems have been brought into equilibrium through ecosystem redesign and the use of selective chemicals, the market potential for these new products will rapidly shrink. But if farming systems continue to depend on regular interventions with these new products, their market life will be drastically curtailed by resistance; highly specific and selective chemistry is often less durable than the old broad-spectrum products, since resistance is more likely to appear in the target weed, insect or fungus. From an ecosystems perspective, IPM is not the substitution of purchased 'soft' products for purchased chemicals. The most precious resource for insect control, for example, is the wealth of naturally occurring predators and parasites, which should be conserved rather than replaced, and augmented with purchased products only when other intensification efforts have failed. The development of resistance to the biological insecticide 'Bt' in parts of the tropics is evidence, in the opinion of the International Institute for Biological Control's director Jeff Waage, that 'commercial interests got IPM only partly right, and therefore failed' (Waage, 1995). But away from the crop protection sphere, companies are realizing that competence in crop biotechnology can open up completely new business opportunities that are making the category of 'pesticide company' increasingly redundant and are challenging the notions of sustainability in the agrifood system. Whole new markets are being created for preserved identity grains and bio-industrial products such as lubricants, fuels and plastics, as well as pharmaceutical products.

Selling services for information-intensive agriculture

The received wisdom regarding 'green' business is that industries can move to a 'dematerialized' commerce, with cradle-to-grave product stewardship or ultimately a knowledge-based economy based on services rather than products. Sustainable agriculture is often described as 'information-intensive' as opposed to 'input-intensive' conventional agriculture. A pesticide company, realizing that it sold crop protection rather than chemicals, could therefore move into the farm services, crop insurance or performance contracting businesses, and deliver the same end result (healthy crops) with far less material throughput and far less environmental impact.

The possibility that a chemical company could 'extract value from a zero-input scenario' was floated by managers from DowElanco at a conference in 1995. The speakers presented a scenario of a shift from selling crop protection to selling crop production. In this scenario, the existing knowledge and management services, fragmented between an unsustainable force of chemical manufacturers, seed companies, distributors, sales representatives, crop consultants and university extension staff, are vertically consolidated to form a critical mass of expertise, in order to commercialize integrated crop management without a product bias.

What would limit such a transformation? The farm services market could be undervalued and distorted. Companies could lack the tools, expertise, organization or credibility to capture value from that market. Sustainable agriculture in its broad definition might not be a lucrative market for any purchased inputs, be they chemicals *or* information. And, finally, there may be considerable shareholder resistance to companies diversifying into farm services. The extreme competition between chemical manufacturers and retailers leads to marketing practices which depress the perceived value of information services (Wolf, 1995). Agrochemical dealers appear to be better positioned than manufacturers to profit from services in sustainable agriculture, with their field presence, local knowledge and personal contacts with farmers. But dealers are concerned that if they try to charge for these services, farmers will defect to competitors. Services such as pest scouting and soil testing are offered free or below cost in order to retain important customers, which makes farmers unwilling to pay independent consultants the full price. The lack of willingness by farmers to pay for IPM-type services has been confirmed in several surveys (see Padgitt et al, 1990). Another major constraint, at least in the US midwest, is performance guarantees which chemical companies offer with their products. Farmers can go back to the dealer for a respray if their fields are not clean after herbicides are applied, and therefore see little incentive to seek out service for a more integrated pest management strategy.

One out of every six acres of US farmland benefits from the services of an independent crop consultant, and these consultants could be a leverage point for change. But few of these consultants are currently offering services in redesigning farm operations for regenerative high-yield agriculture, or even offering significant reductions in pesticide use. Pesticides are very cheap relative to their benefits, with herbicides giving an average of $7 to $15 return for each dollar spent. Most crop consultants place much more emphasis on production rather than cost reduction. The price of a faulty 'no spray' recommendation – damage to crop and reputation – drives consultants to err on the conservative side. Many consultants seem to act as filters for processing the mass of crop protection information into simple, profitable recommendations. The magazine of the National Association of Independent Crop Consultants in the US is as full of agrochemical advertisements as any other farm publication. In fact, surveys have shown that farmers who hire professional crop consultants use on average *more* pesticides than those who do not employ these services. A survey of Nebraska farmers showed that the use of consultants for crop production information declined from 37 per cent with the least sustainable cluster to 0 to 8 per cent with the most sustainable segment (Bernhardt and Allen, 1994).

So far, money spent for local consulting services tends to circulate and remain in farming communities. A move by chemical companies into the service–information business may not be judged as pro-sustainability, if the profits generated by those services leave the community in the same way as expenditures on chemical inputs. An alternative approach would be some kind of shared ownership and shared control between farmers and companies in the development and marketing of products and services. Then there is the question of trust. Farmers will be deeply suspicious of the vested interests and agenda of pesticide companies working as consultants just to make extra sales. 'Given the choice, most farmers prefer to have their recommendations made by someone whose paycheck isn't based on how much product they sell.'

Pesticide companies and their seed subsidiaries have highly centralized research infrastructures to screen chemicals for a limited spectrum of physical and biological characteristics, aimed at developing durable products with market potential over a wide geographical range. These products are sold with relatively limited field support from the manufacturers. Sustainable and biologically intensive agriculture is, by contrast, a local endeavour. A service company or franchise that has ambitions beyond crop scouting and soil sampling requires the support of farming systems-research which takes into account the positive interactions between landscape, soils, crops, pest-suppressing organisms and techniques, livestock and markets for maximizing profitability and minimizing expenditure on purchased inputs. Pesticide company research and outreach would clearly have to completely reengineered to provide this blend of local service and cutting-edge technology; the same can be said of public extension services (Keeney and Vorley, 1997). This shows that without reengineering, there is a danger that companies which seek to diversify into services for sustainable agriculture will package the skills and knowledge of sustainable farming into a product and deliver the technology in the same way as chemical inputs have been sold for the past three decades. Advocates of sustainable agriculture warn strongly against this cargo cult of purchased technologies and expertise (Röling, 1993; Ehrenfeld, 1993, p170), proposing instead that agricultural development is built on participation with farmers and integrates local knowledge – although these concepts have advanced much further in developing countries than in the North (Pretty, 1995; Matteson, 1996).

However, sustainable agriculture risks demanding too much of farmers at a time of increased dependence on off-farm employment. This divergence between meeting the expectations of the land and meeting the expectations of farmers could be bridged through a new breed of advisory services that have a facilitating role, working with farmers to learn, develop technology, and become expert at applying general principles to their situation. The most successful of the new breed of advisors, such as Bob Cantisano in California, work by helping farmers to rethink their fertility and pest management problems. 'The reason farmers hire me', says Cantisano, 'is that I might cut six months or a year off the learning curve and help save money at the same time' (Klinkenborg, 1996). Another constraint on diversification into farm services, which Koechlin and Wittke draw attention to in Chapter 5, is that shareholders will not readily accept a decrease in the earning power of a company for a diversification they can effect more cheaply by buying shares from different companies.

We conclude that if the only new business options that companies seek in sustainable agriculture are substitution technologies – softer biological, chemical or seed-based, pest-control solutions promoted alongside the old 'hard' chemical stalwarts – then the fundamental principles of sustainability have been overlooked and an opportunity for redesign has been missed. But a 'simple' shift from selling chemicals to selling sustainability services is also fraught with difficulty and risks achieving neither the ecological benefits nor farmer acceptance (and hence profitability) that some business commentators have predicted.

Of course, some of the unattractiveness of new businesses in sustainable agriculture stems from the defects inherent in the market for crop-protection products and services. Products with negative externalities (costs to a third party not reflected in the product price) are priced below their real cost to society, and farmers who adopt more sustainable approaches are not operating on an economically level playing field. That is why considerable effort has been made in this project to understand the 'true' price of pesticides, to explore the potential for market corrections (perhaps in the form of green taxes) and to drive more sustainable innovations by allowing ecologically rational consumer choices. 'Green taxes on energy and chemicals', says Paul Hawken, 'will reverse the disincentives to use sustainable methods of agriculture, and will promote widespread use of existing benign methods to control pests and increase yields, while providing existing chemical companies with incentives to go back to the drawing board and invent farming techniques that improve yield and enhance life' (Hawken, 1993, p186). This statement deserves critical scrutiny.

MARKET CORRECTION AND TECHNOLOGICAL SUSTAINABILITY

The survey by Foster and colleagues in Chapter 4, grounded in the theory outlined by Pearce and Tinch in Chapter 3, obtained estimates for the social costs associated with different dimensions of the impacts of pesticides in cereal farming – their effects on wildlife and human health. The figures indicate that society's willingness to pay (WTP) to reverse the decline in a species of farmland birds is considerably higher than their WTP to avoid a single case of mild human illness. This work shows the importance of looking beyond the simple anthropocentric surveys of WTP to avoid personal health risks from pesticide residues on food; the public recognizes *environmental* health as an important attribute of their food supply. We should note that the decline in farmland bird species is associated with the general intensification of agriculture, of which pesticide use is one component. Foster et al used their survey data to estimate how much cereal pesticides would have to be taxed in order to bring their market price in line with these externalities. Their conservative estimate – a 60 per cent ecotax – is a wake-up call for the industry. And when the tax is used not as a uniform tax, but as a means to differentiate between hard and soft chemicals, Foster et al clearly show how a differentiated tax could positively influence farmers' decision-making in selecting more ecologically benign technologies.

Environmental-economics research demonstrates its considerable value as a *tool* in the transition to more sustainable technologies. As Thompson (1995, p117) writes:

> *Economics... does not in itself complete the reconstruction of stewardship to create an adequate philosophy of agriculture. Yet economic approaches offer far more than detractors admit... Resource economics is capable of showing why things go wrong when markets fail to provide incentives to bring about outcomes that everyone wants... It is capable of providing insight into many values that are routinely omitted from decision-making in agriculture. The values measured by resource economists are incomplete, but measurement of them can be helpful both in understanding and in arguing for environmental policies.*

Environmental economics has its fair share of detractors. They accuse it of collapsing a wide range of factors into a single index such as cash or welfare, of ignoring people's moral and cultural attitudes (including altruism), of failing to distinguish between the perceived values of species or landscapes and their core values to the working of ecosystems, and most importantly of recommending taxes which are 'socially optimal' but fail to change behaviour and meet environmental objectives. A school of thought known as *ecological economics* stresses that rather than try to economize ecology, we should try instead to ecologize the economy (Colby, 1990).

It is clear that unsustainability and the development of bad solutions is not just a question of market failure. It is also a question of policy failure in its literal sense – a failure to balance interests, leading to an imbalance of priorities and the development of 'solutions' which benefit only one sector of society (and are likely to be curative rather than preventative, and short-term rather than durable). Failure to address these imbalances will make implementation of the polluter pays concept and ecotaxation political non-starters in all countries with powerful agribusiness lobbies.

> *If polluters under prevailing economic and social conditions do not pay, and governments are invoked to make them do so, how governments proceed with this task will depend on the political and social contract between governors and governed, on balances of rights and responsibilities and the institutions that express and enforce them.*

The OECD's director for the environment, Bill Long, noted ruefully that: 'while it is often easy to gain everyone's agreement for change at the conceptual level, it is another matter entirely to effect change. As ministers know better than I, the potential losers from change inevitably show up. If they happen to represent strong special interest groups, implementing change is daunting at best; impossible at worst.' Long quotes an environment minister from an OECD country who said that: every time he tried to pursue his government's commitments to sustainable development by introducing changes, whether they be economic instruments to internalize environmental costs, eco-labels, or removal of environmentally damaging subsidies, he was labelled either a destroyer of national competitiveness by economic interests at home, or a 'green protectionist' abroad.

Government influence is being overshadowed by international markets and trade regulations to the extent that unilateral government action – in setting ecotaxes on toxic pesticides, for example – is less feasible. It is naive to expect companies acting rationally in conditions of such distorted balance of power to agree to large ecotaxes on profitable pesticides in the name of long-term sustainability. However strongly a CEO agrees with the concept of market correction and ecotaxation, the company and its customers are likely to object strongly to a substantial tax, feeling that they are being charged twice: first when they meet the regulatory standards (a very great expense in the case of pesticide development), and second when they pay the taxes (Cairncross, 1995, p65). Considering these political constraints as 'letting prices speak the truth', we feel that environmental economic research on pesticides is best used by companies and regulators as a decision aid in environmental management. It can assist companies in weighing technological options, which products to invest in, and which to withdraw from the market. We recommend that companies, academics and regulators collaborate on more detailed research to evaluate the true costs of pesticides, and to differentiate 'soft' chemicals from those with high environmental and health costs.

Environmental economics is not a substitute for the political and social dialogue which is required to set societal objectives (Redclift, 1996, p115). As Thompson (1995, p117) writes, 'Ecological economics will extend the contribution of economic thinking... but as the ecological economists themselves admit, it is no substitute for ethics. We must still decide our priorities and goals the old-fashioned way, through deliberation, debate and argument.' Take the furore over genetically engineered grains and oilseeds entering food systems in the EU. The failure of companies such as Monsanto and Novartis to embed these supposedly 'more sustainable' technologies into the social and political framework has brought them disrepute rather than acclaim. Industry's mishandling of biotechnology has brought the vague theoretical ideas of stakeholder involvement, called for in Agenda 21 and alluded to by the Business Council for Sustainable Development, suddenly into sharp relief.

NEW STAKEHOLDER PARTICIPATION AND INVOLVEMENT

We have noted in Chapter 1 that both industry and Agenda 21 have called for a much deeper level of stakeholder participation in setting (or at least approving) the agenda for business and technology development (see Chapter 1, Table 1.4). Stakeholder, positive-sum perspectives are part and parcel of the ecosystems approach, including that applied by the International Joint Commission in the Great Lakes (EPA et al, 1995). Calls for a closer relationship with stakeholders acknowledges the failure of most attempts at communication between business, the public and special interest groups, which have usually turned out to be attempts at recruitment (see Chapter 6). But even when the decision has been made to open up the strategy-making process to a two-way dialogue with a wider range of stakes, there remain the issues of which stakeholders are to be involved and on whose terms.

Industry, science and government have viewed the lodestar of sustainability in highly managerial and technocratic terms, defined by experts from the natural sciences or economics (Orr, 1992). There is a very serious risk that if industry or governments invite wider societal participation in forums dominated by this technical 'expert' knowledge, the result will be grassroots antipathy rather than a broad consensus (Grove-White, 1996). In Chapter 6, Mayhew and Alessi point out the dangers of an exclusive coalition between official institutions ('the system') and science that excludes the powerful strain of opinion (labelled 'sustenance') which looks beyond the narrowly defined physical hazards of individual pesticides or toxics. This 'sustenance' grouping represents the *social* movement of environmentalism, marked by a mistrust of mainstream institutions, such as government and industry and its culture of 'expert' knowledge. It expresses concern not only about the damaging impacts of specific industrial practices, but also the cultural tensions surrounding the values and ways of thinking that got us into this predicament in the first place (Grove-White, 1996). Furthermore, industries and policy-makers that forge ahead with an agenda for sustainable agriculture or green business, narrowly grounded in natural resource issues, are risking talking at cross purposes with other sectors of society by failing to ask *what* is being sustained, for *how long*, and for *whose benefit*. The risks of taking too narrow an approach are laid bare when the interpretations of agricultural 'sustainability' of scientific faculty and farmers are compared. A study by Dunlap et al (1992) clearly demonstrated that university faculty tend to define sustainability in ways which emphasize environmental protection and resource conservation, while farmers have a broader definition that puts emphasis on socio-economic factors, especially the survival and well-being of rural residents and farm communities.

Of course, acknowledging the social aspects of sustainability will be uncomfortable for pesticide producers. Pesticides – especially herbicides – have been part of a suite of technologies that have released millions of farmers and farmworkers from the land. At a time when the value of horticulture, agriculture and nature in the rehabilitation of culture and community is beginning to be appreciated – a time when we appear to need *more* rather than fewer people on the land – there will be a difficult reevaluation of the true meanings and interpretation of efficiency, productivity and convenience. But industry and policy-makers must get participation and stakeholder involvement moving and on the right footing. We have seen in Chapter 7 that participatory futures methodologies have excellent potential in this regard, not necessarily to know the future but to make better decisions today.

Visioning and the application of futures methodology

The future search technique which was applied to the pesticide industry–sustainability question by Morley and Franklin (Chapter 7) allows a rich and productive interaction between stakeholders through its assumptions and its methodology (Weisbord, 1992, pp

3–16; Emery, 1992). The approach assumes that people are purposeful and can be ideal-seeking. It assumes that people want to learn and create their own future. Future search allows participatory, self-managed planning and policy-making among people with diverse interests by working backwards from the preferred future. It focuses awareness away from negative stereotypes and barriers of culture, personality, power, status and politics, towards shared goals. It seeks to hear and appreciate differences, not to reconcile them.

It is astonishing to realize how quickly the future search conference described by Morley and Franklin homed in on leverage points in the pesticides and agricultural sustainability issue, once the discussion had moved beyond the usual string of agricultural woes. As we have seen at various points in the course of this book, many of those leverage points are not found where the farmer or farmworker uses pesticides, or inside the gates of specific pesticide companies, or in the corridors of policy-makers and regulatory agencies. Note the list of opportunities identified in the search in Chapter 7. Four of the nine topics involve moving the goal posts at the marketing and consumption end of the agrifood chain through awareness-building or strengthened communities. Two more address other areas of public action: a need for greater democratic control over technology choice, and mechanisms for building trust among diverse stakeholders. An opportunity for business to develop a stewardship ethic received so many votes for the final 'tasks/actions' stage of the search, that two groups were formed to tackle this issue. Only one area identified – watershed management – was primarily directed at farmers.

The power of future search as a bottom-up visioning and planning tool stands in stark contrast to many of the top-down visions foisted on employees by born-again CEO environmentalists. A vision may be spread through a company by training sessions in which employees jump through hoops until they get it. But that is no way to build the literacy, ownership and unity of purpose necessary for the kind of corporate change required. It does not progress beyond the early win–wins of ecoefficiency towards the deeper redesign of an organization. For a pesticide company or a regulatory agency, it would be a valuable exercise to sponsor three search conferences – one comprised only of company employees (the whole company system), one with stakeholders from the entire agrifood chain, and one for company shareholders. To avoid the problems of too narrow an interpretation of sustainability, and too narrow a view of the company's mission (how can pesticides be used in sustainable agriculture?), all search conferences would be asked to envision a perfect food system, from farm and rural landscape to table. The food system is one in which we are all truly 'insiders' (see CISA, 1995; Anderson, 1995).

Through these search processes, companies are brought face-to-face with other ways of knowing; and environmentalists come face-to-face with constraints on companies, especially their need to protect shareholder value. Consumers come face-to-face with the fact that their choices have a direct impact on farmers' practices in terms of technology choice and landscape features. They are confronted with the fact that much of the pesticide use in developing countries is an extension of the North's ecological footprint, in its demands for feed grains for meat production, out-of-season fruits and vegetables, and cotton fibre and clothing. All stakeholders learn to appreciate that the pesticide issue is not just about pesticides, but is symptomatic of a society 'stuck with ways of knowing, organizing, valuing and doing things, with tightly intertwined roots of unsustainability' (Norgaard, 1992).

The term sustainability should be avoided in search conferences, as it lacks public resonance and sidetracks debate into the minefields of definitions. One participant at the future search conference remarked after two days that 'to promote dialogue, we must ask: what kind of future do you want? rather than: how are we going to get a sustainable agriculture?' Based on evidence that the personal moral and ethical positions of company employees, policy-makers and environmental activists are much closer than their organizational rhetoric would have us believe (see Cairncross, 1995, p187; Glasser, et al 1994), we hypothesize that searches involving company employees, shareholders or diverse food-chain stakeholders would arrive at very similar conclusions and recommendations. These searches could form the core of a very powerful instrument for change, regularly reconvening to review progress in implementing their plans. We recommend that future search or related techniques should be used as a long-term planning and evaluation tool for catalyzing and planning institutional change.

Technology choice

Once a common vision of a preferred future has been agreed upon, company research managers and policy-makers have a much clearer view of the kinds of products and services which will be needed to achieve that preferred future. Until now, research managers have been struggling to hedge their bets from a shrinking R&D purse in the high-stakes lotteries of chemistry, biotechnology and seeds. How much should each receive? How will the market for biotechnology develop, and how will it affect the demand for new chemicals? Instead, we propose that research, and especially development, should be driven less by the scientific potential of the technologies and more by a common purpose and vision of what the agrifood system should look like in a more sustainable world.

As usual, the parallels between business and agriculture are astonishing. In an ecosystem-based agricultural management strategy termed holistic resource management (HRM) by its founder, Allan Savory, the first step to sustainable agriculture must be a goal which is based on a quality of life statement that all parties support (Savory, 1988, p445). A form of production required to support that quality of life is then agreed upon, which in turn determines the type of landscape that will support the production and sustain the quality of life. Likewise, a company that sets a goal based on a preferred future that all parties (public, shareholders, employees) support will see quite easily what kind of technologies and services are required to achieve that future, and what kind of organization is best suited to develop and sell those products and services. A company's sole objectives to be number one, or maximize returns to shareholders, or to grow in size each year, are as much 'non-goals' as a farmer's sole aim to become more efficient. Doing things right (efficiency) must be nested within a larger frame of doing the right thing, which Drucker (1973) calls effectiveness.

Our intention is not to abdicate all control of company strategy to a multistakeholder group. There is clearly a role for speculative and basic research in the private sector. But companies in a more sustainable world cannot afford to base technology choice simply on a narrow win–win scenario: profit and conservation of natural resources. The experiences with biotechnology highlight the risks of such a course. The developers of crop biotechnology have stepped into the same quagmire of social and ecological revenge effects as in the chemical era of pest control, thanks to the scientists' intoxication with technology's potential and their obliviousness to the broader concerns and latent tensions within the public at large:

> *In the biotechnology case, the knowledge culture.. makes the tacit assumption that people care predominantly about narrowly defined, highly specific physical hazards, rather than the broader social*

relations surrounding the manipulation and marketing of life forms.

Robin Grove-White (1996)

Several key questions remain concerning future search applications to business strategy. Empirical evidence is required to demonstrate to companies that they can do right by doing the right thing. Can a company afford to unilaterally start setting its strategy according to a *preferred* future rather than from forecasts of likely or probable futures? Is any company – even a giant multinational chemical company – big enough to influence the future rather than just anticipate it? Is this technique suitable for the niche marketing of garden tools but more akin to commercial suicide for a chemical company?

We have to start somewhere, and the individual corporation is probably a better place to start than either national government or industry associations. This reflects Handy's 'second curve thinking', in which a company or organization is forced to challenge the assumptions underlying its present strength by asking, 'If we could reinvent ourselves, what would we look like?' (Handy, 1994, p56). Systems thinking and future-oriented planning tools, when used by a literate and collegial organization, can provide a conceptual platform from which we can become active participants in shaping our reality, 'from reacting to the present to creating the future' (Senge, 1990, p69).

FIRST STEPS

Institutional shifts are not clear cut; rather, they may be messy affairs in which new institutions – perhaps initially developed out at the margins of the organization – are built into old models. We can expect that a pesticide company in transition towards sustainable agriculture will be rife with double standards and mixed signals, and that a significant number of employees will continue to experience tension between their personal values and the prevailing policies of their institution.

Companies continue to sell hazardous products, in apparent contradiction to their vision for sustainable development. They continue to offer sales incentives to their field forces, to offer performance guarantees with their herbicides, and to advertize in ways which play on farmers' fears of crop loss. Despite the rhetoric and assurances, companies keep these products and practices alive to pay for R&D in the hope of discovering chemical and genetic substitutes. But in order to build the levels of credibility and innovation which a transition to sustainable agriculture demands, companies must take decisive first steps to deal with their most glaring nonvisionary products and marketing practices, and also to legitimize the personal values of their employees.

Strong action from top management is essential to clear out undesirable products from a company's portfolio. Bad products may be trivial to the company as a whole – after all, CFCs represented only 2 per cent of DuPont's total revenues – but may seem critical to the survival of devolved business units. Consequently, the first steps away from greenwash and towards carrying out a vision for sustainable development must be resolute and proactive, and demonstrate a genuine ability to self-regulate.

Redesigning an industry or a government agency will require the engagement of employees who find personal fulfilment in their work, and who find the institution's case legitimate (Oehl, 1992); in other words, the purpose of the institution should echo personal purpose (Janger and Edmonson, 1990). Simply maximizing shareholder value alienates the 'knowledge workers' whom the company or agency will depend upon for successful transformation (Drucker, 1993, p80). Ciba's Alex Krauer told *Der Speigel* (2 October 1989) that: 'Young women and men who are starting their careers with us have values radically different from those of 20 years ago. They ask critical questions. If their expectations are out of tune with our behaviour, we have motivational problems.' Monsanto's previous chairman, Richard Mahoney, directed his strategy of drastic reductions in emissions in part at his own workforce who, he argued, 'share the background and values of many members of environmental groups; some may well be members themselves' (Cairncross, 1995, p187). So what are we to feel when a survey of ICI Agrochemicals (now Zeneca) employees reveals that only 30 per cent consider pesticides to be essential for food production (Otter, 1992), or when many policy advisors to European governments admit suppressing deeply held ecological values to satisfy institutional constraints (Glasser et al, 1994)? We propose that companies and regulatory agencies put a far greater emphasis on the legitimization of their employees' personal values.

CLOSING REMARK – NEW BUSINESS OR NEW LEARNING PROCESS?

We have sought leverage points to support a complementary rather than antagonistic role for the pesticide industry in sustainable agriculture, using the industry's own language. The language is there, in business declarations and corporate visions for sustainable development, and the leverage points do exist which can move the debate out of the mire of orthodoxy and defensiveness.

Two key ingredients are missing. The first is literacy in the language of sustainability, through the ranks of managers in companies and government agencies. Perspectives of sustainability as a functioning ecosystem, and of sustainability as the regeneration of community, are needed. The second criterion is related to the first, to focus on ends rather than means. We need a common vision, agreed on by all company or agency stakeholders, of what a preferred future looks like, to guide decisions and choices today. A top-down vision for sustainable development is a contradiction in terms. And a vision in an organization that does not have a culture of continual learning is doomed to remain as a set of principles rather than as a catalyst for innovation.

We as a society have declared war on pests and are fashioning ever more sophisticated swords from the far reaches of industrial technology and bio-engineering, under the banners of ending hunger and of improving efficiency. But the farmers who work with, rather than against, the biological basis of agriculture see folly in this separation of pests from the rest of the farming system. Those farmers have a great deal in common with systems thinkers in business, who see that 'you and the cause of your problems are part of a single system'. Industry should perhaps look first to the philosophies that inspired regenerative forms of agriculture, namely holism and systems thinking, before searching for new revenues in products and services in 'sustainable' agriculture. This is first and foremost an exercise in enlightenment, in literacy, in challenging the perceptions and world views that have given rise to the pest wars ethic. The agricultural sector which the pesticide industry has set out to modernize may be the richest source of technical and managerial innovation.

Perhaps at the end of such an exercise there are very few immediate spin-offs in the form of products for sustainable agriculture. The ploughshares may be made by a completely different enterprise than the swords. But the changes in management thinking that accompany this new view of agriculture could have a hugely beneficial influence throughout any company.

References

Altieri MA, Nicholls CI and Wolfe MS (1996) 'Biodiversity – a central concept in organic agriculture: restraining pests and diseases', pp 91–112, in Østergaard TV (ed) *Fundamentals of Organic Agriculture*. Proc. 11th IFOAM International Scientific Conference, 11–15 August 1996, Copenhagen, Vol 1. IFOAM Tholey-Theley, Germany.

Anderson MD (1995) 'The life cycle of alternative agricultural research'. *American Journal of Alternative Agriculture*, 10(1), pp 3–9.

Avery D (1995) *Saving the planet with pesticides and plastic.* Hudson Institute, Indianapolis.

Benbrook C (1996) *Pest Management at the Crossroads.* Consumers Union, Yonkers, NY.

Bernhardt K and Allen JC (1994) 'Adoption/diffusion of sustainable agricultural practices: what influences change?' Paper presented at the Annual Rural Sociology Society Meeting, 10–14 August 1994, Portland, OR.

Berry W (1978) *The Unsettling of America: culture and agriculture.* Sierra Club Books, New York.

Berry W (1981) *The Gift of Good Land: further essays cultural and agricultural.* North Point Press, New York.

Bromley DW (1991) 'Technology, technical change, and public policy: the need for collective decisions'. *Choices*, Second Quarter, pp 5–13.

Burnside K and Hoefer R (1995) Unpublished paper presented at the conference Privatization of Technology Transfer in US Agriculture: Research and Policy Implications, 25–26 October 1995, University of Wisconsin-Madison.

Cairncross F (1995) *Green, Inc.* Earthscan and Island Press, London and Washington, DC.

CISA (1995) *Sustainable Food and Farming in the Connecticut River Valley: a vision.* Community Involved in Sustainable Agriculture (CISA) Future Search Conference, 16–18 March 1995. SearchNet, 433 Kelly Drive, Philadelphia, PA.

Clayton AMH and Radcliffe NJ (1996) *Sustainability: a systems approach.* Earthscan, London.

Colby ME (1990) 'Environmental management in development: the evolution of paradigms'. Discussion paper #80, World Bank, Washington, DC.

Coppolino EF (1994) 'Pandora's poison'. *Sierra*, September/October 1994, pp 41–77.

de Bono E (1970) *Lateral Thinking: creativity step by step.* Harper & Row, New York.

Dinham B (1993) *The Pesticide Hazard: a global health and environmental audit.* Zed Books, London and New Jersey.

Drucker PF (1973) *Management – Tasks, Responsibilities, Practices.* Harper & Row, New York.

Drucker PF (1993) *Post-capitalist Society.* HarperCollins, New York.

Dunlap RE, Beus CE, Howell RE and Waud J (1992) 'What is sustainable agriculture? An empirical examination of faculty and farmer definitions'. *Journal of Sustainable Agriculture* 3(1), pp 5–36.

Ehrenfeld D (1993) *Beginning Again: people and nature in the new millennium.* Oxford University Press, New York.

Ekins P (1995) 'Economics and sustainability'. pp161–204, in Ravaioli C *Economics and the Environment.* Zed Books, London and Atlantic Highlands, NJ.

Ellyard P (1994) 'A clean future. Creating a sustainable future. Trade and the future environmental world'. Addresses to the 10th IFOAM International Organic Agriculture Conference, 11–16 December 1994, Lincoln University, New Zealand.

Emery M (1992) 'Training search conference managers', pp 325–342, in Weisbord et al (1992).

EPA et al (1995) *Practical steps to implement an ecosystem approach in Great Lakes management.* US Environmental Protection Agency, Environment Canada, International Joint Commission, Wayne State University.

Ghersa CM, Roush ML, Radosevich SR and Cordray SM (1994) 'Coevolution of agroecosystems and weed management'. *BioScience*, 44(2) pp 85–94.

Glasser H, Craig PP and Kempton W (1994) 'Ethics and values in environmental policy: the said and the UNCED', pp 83–107, in van den Burg JCJM and van der Straaten J (eds) *Toward Sustainable development: concepts, methods, and policy*. Island Press, Washington, DC.

Grove-White R (1996) 'Environmental knowledge and public policy needs: on humanising the research agenda', pp 269–286, in Lash S et al (eds) *Risk, Environment & Modernity: towards a new ecology*. Sage Publications Ltd, London.

Hamilton ND (1996) 'Tending the seeds: the emergence of a new agriculture in the United States'. *Drake Journal of Agriculural Law*, 1(1) pp 7–29.

Handy C (1994) *The Empty Raincoat: making sense of the future*. Hutchinson, London.

Hartig JH and Vallentyne JR (1989) 'Use of an ecosystem approach to restore degraded areas of the Great Lakes'. *Ambio*, 18, pp 423–28.

Hawken P (1993) *The Ecology of Commerce: a declaration of sustainability*. HarperBusiness, New York.

Hill SB and Ramsey J (1977) 'Weeds as indicators of soil conditions'. *The MacDonald Journal*, June 1977, pp 8–12.

Howard A (1940) *An Agricultural Testament*. Oxford University Press, London.

Janger SM and Edmonson AC (1990) 'Business in transformation', pp161–195, in Harman W and Hormann J *Creative Work: The constructive role of business in transforming society*. Knowledge Systems Inc, Indianapolis, IN.

Keeney D and Vorley W (1997) 'Can privatization of information meet the goals of sustainable agriculture?' in Wolf S (ed) *Privatization of Technology and Information Transfer in US Agriculture*. St Lucie Press and Soil and Water Conservation Society, Ankeny, Iowa, (in press).

Kellert SR and Wilson EO (eds) (1993) *The Biophilia Hypothesis*. Island Press, Washington, DC.

Klinkenborg V (1996) 'Amigo Cantisano's organic dream'. *New York Times Magazine*, 10 March 1996, pp 48–54.

Kloppenburg J Jr, Hassanein N and Burrows B (1996) *Does technology know where it's going? (12 reasons to stop expecting modern biotechnology to create a sustainable agriculture and what to do after the expectation has ceased)*. The Edmonds Institute, Edmonds, Washington, DC.

Lenzer R and Upbin B (1997) 'Monsanto v. Malthus'. *Forbes* 10 March 1997, pp 58–64.

Lewis CA (1996) *Green Nature/Human Nature: the meaning of plants in our lives*. University of Illinois Press, Urbana and Chicago.

MacRae RJ, Henning J and Hill SB (1993) 'Strategies to overcome barriers to the development of sustainable agriculture in Canada: the role of agribusiness'. *Journal of Agricultural and Environmental Ethics*, pp 21–51.

Magretta J (1997) 'Growth through sustainability: an interview with Monsanto's CEO, Robert B Shapiro'. *Harvard Business Review*, January–February 1997, pp 79–88.

Matteson PC (1996) 'Implementing IPM: policy and institutional revolution'. *Journal of Agricultural Entomology*, 13(3), pp 173–83.

Montague P (1993) 'How we got here—part 2: who will take responsibility for PCBs?' *Rachel's Hazardous Waste News*, 329 (18 March 1993), Environmental Research Foundation, Annapolis, MD.

Mueller K and Koechlin D (1992) 'Environmentally conscious management', pp 33–60 in Koechlin D and Müller K (eds) *Green Business Opportunities: the profit potential*. Pitman, London.

Norgaard RB (1992) 'Sustainability: the paradigmatic challenge to agricultural economists' in Peters GH and Staunton BE (eds) *Sustainable agricultural development: the role of international cooperation*, pp 92–101. Proceedings of the 21st International Conference of Agricultural

Economists, Tokyo Japan, 22–29 Aug 91. Dartmouth, Brookfield, Vermont.

Oehl TP (1992) 'Thoughts about the changeability of corporate cultures', pp 207–225, in Koechlin D and Müller K (eds) *Green business opportunities: the profit potential.* Financial Times/Pitman, London.

Orr DW (1992) *Ecological Literacy: education and the transition to a postmodern world.* State University of New York Press, Albany, NY.

Otter J (1992) 'Some aspects of environmental management within a chemical corporation', pp 81–98, in Koechlin D and Müller K (eds) *Green business opportunities: the profit potential.* Financial Times/Pitman, London.

Padgitt S, Watkins S and Grundman D (1990) *Assessing extension opportunities for Integrated Pest Management.* Iowa State University Publication, IFM13, Ames, IA.

Pretty JN (1995) *Regenerating Agriculture: Policies and practice for sustainability and self-reliance.* Earthscan, London.

Redclift M (1996) *Wasted: counting the costs of global consumption.* Earthscan, London.

Regenstein L (1986) *How to Survive in America the Poisoned.* Acropolis Books, Washington, DC.

Richmond B (1994) 'System dynamics/systems thinking: let's just get on with it'. Paper presented at the 1994 International Systems Dynamics Conference, Sterling, Scotland, 1994. Available at http://www.hps-inc.com/st/paper.html.

Ripper WE (1994) 'Biological control as a supplement to chemical control of insect pests'. *Nature,* 153, pp 448–52.

Robinson RA (1996) *Return to Resistance: breeding crops to reduce pesticide dependence.* agAccess, Davis, CA.

Röling N (1993) 'Agricultural knowledge and environmental regulation in the Netherlands: a case study of the Crop Protection Plan'. Sociologia Ruralis, 33(2), pp 261–280.

Sachs W (1993) 'Global ecology and the shadow of 'development'', pp 3–21, in Sachs W (ed) *Global Ecology: a new arena of political conflict.* Zed Books, London and New Jersey.

Savory A (1988) *Holistic Resource Management.* Island Press, Washington, DC.

Schönbeck F (1989) 'Pflanzengesundheit – eine Herausforderung an den Pflanzenschutz'. *Nachrichtenbl. Deut. Pflanzenschutzd,* 41(12), pp 204–7.

Senge PM (1990) *The Fifth Discipline: The Art and Practice of the Learning Organization.* Currency-Doubleday, NY.

Solow R (1993) 'Sustainability: an economist's perspective', pp 179–87, in Dorfman R and Dorfman N (eds) *Economics of the Environment* (third edition). WW Norton & Co, NY.

Tenner E (1996) *Why Things Bite Back: technology and the revenge of unintended consequences.* Alfred A Knopf, NY.

Thompson PB (1995) *The Spirit of the Soil: agriculture and environmental ethics.* Routledge, London & New York.

Thrupp LA (1991) 'Sterilization of workers from pesticide exposure: the causes and consequences of DBCP-induced damage in Costa Rica and beyond'. *International Journal of Health Services,* 21(4), pp 731–757.

Thrupp LA (1996) *New Partnerships for Sustainable Agriculture.* World Resources Institute, Washington, DC

Turner RK, Pearce D and Bateman I (1994) *Environmental Economics: an elementary introduction.* Harvester Wheatsheaf, Hemel Hempstead, UK.

van den Bosch R (1978) *The Pesticide Conspiracy.* Doubleday, Garden City, NY.

van Emden HF (1991) 'The role of host plant resistance in insect pest mis-management'. *Bulletin of Entomological Research,* 81, pp 123–126.

Waage J (1995) 'Divergent perspectives on the future of IPM'. Paper presented at the

International Food Policy Research Institute (IFPRI) workshop: Pest management, food security, and the environment: the future to 2020. Washington, DC, 10–11 May 1995.

Waage J (1996) 'Integrated Pest Management and biotechnology: an analysis of their potential for integration', pp 37–60, in Persley GJ (ed) *Biotechnology and Integrated Pest Management.* CAB International, Wallingford, UK.

Waltner-Toews D (1996) 'Ecosystem health – a framework for implementing sustainability in agriculture'. *BioScience*, 46(9), pp 686–89.

Watt KEF (1970) 'The systems point of view in pest management', pp 71–83, in Rabb RL and Guthrie FE (eds) *Concepts of Pest Management.* North Carolina State University, Raleigh, NC.

Weisbord MR et al (1992) *Discovering Common Ground: how future search conferences bring people together to achieve breakthrough innovation, empowerment, shared vision, and collaborative action.* Berrett-Koehler, San Francisco.

Welford R (1994) *Cases in Environmental management and Business Strategy.* Pitman Publishing, London.

Wigglesworth VB (1945) 'DDT and the balance of nature'. *Atlantic Monthly*, 176(6) pp 107–13.

Wolcott RM (1993) 'Producing environmental quality: an emerging market for agriculture in America'. Paper presented at workshop Reinventing Agriculture, 8–9 December 1993, Chicago, IL.

Wolf S (1995) 'Cropping systems and conservation policy: the roles of agrichemical dealers and independent crop consultants'. *Journal of Soil and Water Conservation*, May–June 1995, pp 263–70.

Contact Addresses

Leopold Center for Sustainable Agriculture
The Leopold Center was established under the Iowa Groundwater Protection Act of 1987 to conduct research into the negative impacts of agricultural practices and to assist in developing and demonstrating alternative practices.

Contact:
Leopold Center for Sustainable Agriculture,
209 Curtiss Hall, Iowa State University,
Ames, Iowa 500111–1050, USA
Tel +1 515 294 3711,
Fax +1 515 294 98696
http://www.leopold.iastate.edu

Centre for Social and Economic Research on the Global Environment (CSERGE)
CSERGE is Britain's leading environmental economics research unit, linking specialists at University College London, the University of East Anglia and a network of advisors and external staff.

Contact:
CSERGE, University College London,
Gower Street,
London, WC1E 6BT, UK
Tel +44 (0)171 380 7874,
Fax +44 (0)171 916 2772
http://www.ues.ac.uk/menu/acad_depts/env/all/resgroup/cserge/noframe.htm

International Institute for Environment and Development (IIED)
IIED was established in 1971 as a policy research institute linking environmental concerns with development needs. The institute advises decision-makers and raises public awareness on such critical issues as sustainable agriculture, tropical forestry, human settlements, drylands management and environmental economics.

Contact:
IIED, 3 Endsleigh Street,
London WC1H 0DD, UK
Tel +44 (0)171 388 2117,
Fax +44 (0)171 388 2826
http://www.ecdpm.org/iied/index.html

Ellipson AG
Ellipson is a Swiss-based consultancy specializing in environmentally conscious management and strategy. Expertise is well developed for life-cycle analysis and environmental reporting.

Contact:
Ellipson AG, Leonhardsgraben 52,
CH-4051 Basel, Switzerland
Tel +41 61 261 9313,
Fax +41 61 601 4641

Economics for the Environment Consultancy Ltd (EFTEC)
ECTEC specializes in environmental economics, providing clients in both the corporate and government sector with economic analysis and quantitative research in order that environmental objectives are fulfilled with maximum economic efficiency.

Contact:
EFTEC Ltd, 16 Percy Street,
London W1P 9FD, UK
Tel +44 (0)171 580 5383,
Fax +44 (0)171 580 5385
Email eftec@eftec.demon.co.uk

ABL Group, University of York
The ABL (Adapting By Learning) based in the Faculty of Environmental Studies at York University, Toronto, specializes in the use of action research in organizational and social change settings.

Contact:
Faculty of Environmental Studies,
York University, 4700 Keele Street,
Toronto, Ontario M3J 1P3, Canada
Tel +1 416 736 2100 x22615,
Fax, +1 416 920 0797
Email dmorley@yorku.ca

A&M Consulting
A&M Consulting provides problem-solving tasks for people in complex social environments. Systems Therapy, the approach developed by A&M, blends the disciplines of Family Systems Therapy, Cultural Anthropology, Human Development and Systems Architectural Design. Rather than simply applying a pre-designed solution, A&M works alongside the client to recast the problem definition and the attempted solutions. Additional services provided by A&M are human systems training, assessment and advice, focus group design and facilitation and systems architecting.

Contact:
198 N Lakeshore Drive,
Glenwood MN 56334, US
Tel +1 320 634 0096
Fax +1 630 435 1782
Email mickmayhew@msn.com or salessi@msn.com

INDEX

Page references in **bold** refer to tables and figures